Our Place in the World Around Us

Environmental Geology Labs

FIRST EDITION

By Elli Pauli

The George Washington University

cognella™
San Diego, CA

Bassim Hamadeh, CEO and Publisher
Michael Simpson, Vice President of Acquisitions
Jamie Giganti, Managing Editor
Jess Busch, Senior Graphic Designer
Brian Fahey, Licensing Specialist

First published in the United States of America in 2015 by Cognella, Inc.

Printed in the United States of America

ISBN: 978-1-63189-094-9 (pbk) / 978-1-63189-095-6 (br)

www.cognella.com 800.200.3908

CONTENTS

ACKNOWLEDGMENTS

While I laid down the groundwork for this publication, the final product is the result of several years worth of collaborating, testing, and revising with the help of a multitude of dedicated people. To begin, I offer my sincerest thanks to Frances Junker—without whom this publication would not have been possible, for her countless hours, alternative suggestions, excellent ideas, and the entertainment she brought to what otherwise would have been a frustrating and tedious bore.

In addition, I owe an immense debt of gratitude to Dr. Catherine Forster and Dr. Giuseppina M. Kysar for giving up hours of their personal time to review this publication and contribute invaluable knowledge and insight.

INTRODUCTION

Environmental geology is the branch of science that teaches humanity how to responsibly use Earth's valuable resources, prevent contamination of its water, air, and soil, and predict its dangerous movements.

Education of the populace on this subject matter has recently become critical to mankind's survival, as Earth's natural equilibrium has been detrimentally disrupted by the demands of a continuously increasing population that exceeds what Earth's systems are designed to support, resulting in escalating environmental problems. This book is designed to make the user aware of the multiple environmental factors that dictate the quality of our lives, and in doing so, educate the user on the healthiest way to co-exist with the world around us.

THE BASICS

OBJECTIVE

This lab is designed to refresh your memory on some basic mathematical and scientific principles that you will need to successfully complete the laboratory course. Use the information in the conversion charts at the end of this lab exercise to answer the following questions.

Part A: Scientific Notation

Recall that we can express very large or very small numbers in scientific notation; this means that we express numbers in terms of powers of 10.

$10^1 = 10$ and $a\star10^1 = a\star10$ (e.g., $3\star10^1 = 30$)
$10^2 = 100$
$10^3 = 1000$

$10^{-1} = 0.1$ and $a\star10^{-1} = a\star0.1$ (e.g., $3\star10^{-1} = 0.3$)
$10^{-2} = 0.01$
$10^{-3} = 0.001$

An important rule is that $10^0 = 1$.

Problems

1. The radius of the Earth is 6371 km. Express this in scientific notation.

   _____

2. A single salt crystal is on the order of 10^{-8} meters in size. Write this number as a decimal.

   _____

Part B: Dimensional Analysis

It's not necessary to memorize every possible conversion that you'll encounter in the course of this class. You can use a succession of conversions to calculate the one that you need, such as in the following example.

Example: How many meters are in a light year (the distance that light travels in one year), given that light travels at a speed of 3.0×10^8 meters per second?

Answer: Recall that Distance = Rate x Time (D=RT). We know the rate (R = 3.0×10^8 m/s) and the time (T=1 year), which leaves distance as our unknown.

Therefore: D = 3.0×10^8 m/s x 1 year

But in order to solve this problem, we must convert years to seconds so the time cancels out leaving only distance.

So how do we calculate how many seconds are in a year? We know the following conversions: 60 secs = 1 min; 60 min = 1 hour; 24 hours = 1 day; and 365 days = 1 year.

$$1 \text{ year x } \frac{365 \text{ days}}{1 \text{ year}} \text{ x } \frac{24 \text{ hours}}{1 \text{ day}} \text{ x } \frac{60 \text{ mins}}{1 \text{ hour}} \text{ x } \frac{60 \text{ secs}}{1 \text{ min}} = 3.1536 \times 10^7 \text{ sec}$$

Now, we can multiply this value by the speed of light to get the distance represented by a light year.

Problems

3.   Finish the above example: how many meters are in a light year? _____ m

4.   Make the following conversions using the conversion factors provided at the back of this lab exercise.

3,168 feet (ft) = _____ inches (in) = _____ miles (mi).

21 meters (m) = _____ kilometers (km) = _____ centimeters (cm).

23,400 seconds (s) = _____ minutes (min) = _____ hours (hrs).

62,300 kg = _____ pounds (lbs) = _____ tons.

220 ft/min = _____ m/s.

2 centuries = _____ seconds.

45 seconds = _____ days.

5.   The Earth is 4.6 billion years old. Given that 1 billion = 10^9, how many centuries is this? *Show your work.* _____ centuries

6.   The Earth's tectonic plates move at rates of 1–10 centimeters per year. How many miles per hour is 5 cm/year? *Show your work.* _____ mph

7.   The United States is approximately 3,000 miles across. How many inches is this? *Show your work.* _____ inches

8.   The islands of Hawaii are volcanic islands built up by layers of successive lava flows. When measured from its base on the ocean floor, one of the Hawaiian Islands stands approximately 9,140 meters. Radiometric dating of the volcanic rocks indicates that the island is 0.9 million years old. What is the average rate of build-up of lava flows in centimeters per year? *Show your work.*

_____ cm/yr

9.   Historical records show that the volcanoes on the island above do not erupt every year, thus in some years, there is no build-up of lava and the height of the island is actually reduced by erosion. What does this say about the actual rate of build-up of the island as compared to the average rate calculated in the previous question?

_____
_____
_____

10.   The Himalaya Mountains were formed by the collision of the Indian subcontinent with Asia. They have the highest rate of uplift in the world (1 cm/yr). If the initial collision between the Indian and Asian subcontinents occurred 40 million years ago, what should the height of the Himalayas be in kilometers (km)? *Show your work.* _____ km

11.   In actuality, the Himalayas are approximately 8.25 km high. Why do you think the actual height of the Himalayas is not the same as the height you calculated in question 10?

_____

12.   Knowing that the actual height of the Himalayas is approximately 8.25 km, what is the net rate of uplift in cm/yr? *Show your work.* _____ cm/yr

Part C: Distance, Area, Volume, Density

In geology, we deal with many spatial relationships. We sometimes need to calculate areas and volumes, and it's important to know how to convert the corresponding units when the calculations are finished.

In the same way that we do dimensional analysis of distances, we can do them for areas or volumes, keeping in mind that the units in the numerators and denominators of our relationships must cancel. For example, we can calculate how many square feet are in one square meter:

$$1 \text{ m}^2 * \frac{(100 \text{ cm})^2}{1 \text{ m}^2} * \frac{(1 \text{ in})^2}{(2.54 \text{ cm})^2} * \frac{(1 \text{ ft})^2}{(12 \text{ in})^2} = 10.7 \text{ ft}^2$$

Note that when we want to square a unit, we also have to square the quantity that goes with the unit. In other words, 1 m^2 is (100 cm)^2 = 1.0 x 10^4 cm^2; *not* 100 cm^2! The same goes for volumes.

Problems

13.   Recall that the area of a rectangle is its length x width. If we pretend that a given country is a rectangle with a length of 2000 km and a width of 500 km, what is its area in km^2? *Show your work.* _____ km^2

14.   What is its area in square miles? *Show your work.* _____ mi²

Use the following chart to help you organize your answers to questions 15–21. Any numbers given to you in questions 15–21 have already been filled in for you.

	Radius	Mass	Volume	Density
Earth	6371 km	5.97 x 10²⁴ kg	_____ km³	_____ kg/m³
Core	_____ km	_____ kg	_____ km³	7.874 kg/m³

15.   Recall that the formula for volume of a sphere is 4/3πr³. If the radius of the Earth is 6371 km, what is its volume in km³? *Show your work.* _____ km³. Record the volume of the Earth in the chart provided above.

16.   The mass of the Earth is 5.97 x 10²⁴ kilograms; if density (ρ) is calculated as ρ = mass/volume, what is the density of the Earth in kg/m³? *Show your work.* _____ kg/m³. Record the density of the Earth in the chart provided above.

17.   What is the mean density of the Earth in g/cm³? *Show your work.* _____ g/cm³

18.   If the Earth's core has a radius equal to half the radius of the Earth, what fraction of the Earth's volume is occupied by the core? (*Hint:* Calculate the volume of the core and divide it by the volume of the Earth.) Record the radius and volume of the core in the chart provided above. *Show your work.* _____

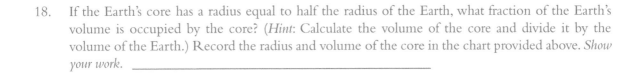

19.   If the density of iron is 7,874 kg/m³, what is the approximate mass of the Earth's core? *Show your work.* _____ kg. Record the mass of the core in the chart provided above.

20.   What fraction of Earth's mass is represented by the mass of the core? *Show your work.*

   _____

21.   What is the mean density of the rest of the Earth? (*Hint:* Find the mass and volume of the rest of the Earth by subtracting the mass and volume of the core from the mass and volume of the Earth, respectively, and use those values to find the density of the mantle + crust.) *Show your work.*
   _____ kg/m³

Part D: Rates and Percentages

22.   Use the Comparative Rates Figure on page 7 to answer the following questions.

a. Fingernails grow at a rate of approximately 15 mm/yr. Which geologic processes are occurring at approximately the same rate? _____

_____

b. Assuming that you can run like an Olympic distance runner (let's say at a rate of about 12.5 miles per hour), do you think that you could outrun:

   i. debris flow? _____

   ii. mudflow? _____

   iii. flowing glacier? _____

Support your answer using mathematics.

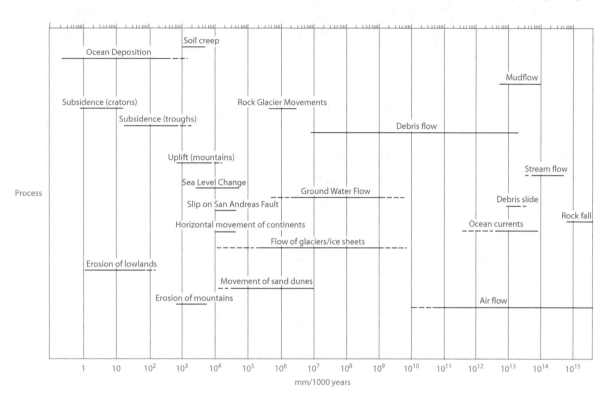

23. The composition of municipal solid waste is approximately 38% paper, 18% yard wastes, 11% metals, 9% plastics, 8% food wastes, 6% glass, 5% wood, rubber, textiles and leather, and 5% miscellaneous inorganic wastes. If 50% of all glass, paper, plastics, and yard wastes were recycled, what percent reduction in waste would be achieved? *Show your work.* _____

24. Assuming that recycling values are as follows: paper, $5/ton; glass, $8/ton; and plastics, $10/ton. Using the values from the previous question, if your local landfill receives 600 tons of refuse per year and they wanted to start a recycling program, would it be most advantageous to start with paper, glass, or plastic? *Show your work.* _____

Conversion Charts

Numerical Value	Verbal equivalent	Prefix	Symbol	Exponential Expression
0.000 000 000 000 000 001	One-quintillionth	atto	a	10^{-18}
0. 000 000 000 000 001	One-quadrillionth	femto	f	10^{-15}
0. 000 000 000 001	One-trillionth	pico	p	10^{-12}
0. 000 000 001	One-billionth	nano	n	10^{-9}
0. 000 001	One-millionth	micro	μ	10^{-6}
0. 001	One-thousandth	milli	m	10^{-3}
0.01	One-hundredth	centi	c	10^{-2}
0.1	One-tenth	deci	d	10^{-1}
1				10^{0}
10	Ten	deca	da	10^{1}
100	Hundred	hecto	h	10^{2}
1000	Thousand	kilo	k	10^{3}
1 000 000	Million	mega	M	10^{6}
1 000 000 000	Billion	giga	G	10^{9}
1 000 000 000 000	Trillion	tera	T	10^{12}
1 000 000 000 000 000	Quadrillion	peta	P	10^{15}
1 000 000 000 000 000 000	Quintillion	exa	E	10^{18}

Linear Measurements	Area Measurements
1 foot = 12 inches	1 mi^2 = 640 acres
1 mile = 5,280 feet	1 acre = 43,560 ft^2
1 nautical mile = 6,076.115 feet	1 acre = 4,840 yd^2
1 km = 1000m = 10^3m; 1 mm = .001m = 10^{-3}m	1 acre = 4,047 m^2

CONVERSION CHARTS

English/Metric and Metric/English Conversions		
	English to Metric	Metric to English
Inches & Millimeters	1 inch = 25.4 millimeters	1 millimeter = 0.0394 inches
Feet & Meters	1 foot = 0.3048 meters	1 meter = 3.281 feet
Yard & Meters	1 yard = 0.9144 meters	1 meter = 1.094 yards
Miles & Kilometers	1 mile = 1.609 kilometers	1 kilometer = 0.6214 miles
in^2 & cm^2	1 in^2 = 6.4516 cm^2	1 cm^2 = 0.155 in^2
ft^2 & m^2	1 ft^2 = 0.0929 m^2	1 m^2 = 10.764 ft^2
yd^2 & m^2	1 yd^2 = 0.836 m^2	1 m^2 = 1.196 yd^2
mi^2 & km^2	1 mi^2 = 2.59 km^2	1 km^2 = 0.3861 mi^2
in^3 & cm^3	1 in^3 = 16.39 cm^3	1 cm^3 = 0.061 in^3
ft^3 & m^3	1 ft^3 = 0.0283 m^3	1 m^3 = 35.3 ft^3
yd^3 & m^3	1 yd^3 = 0.7646 m^3	1 m^3 = 1.308 yd^3
Quarts & Liters	1 quart = 0.946 liter	1 liter = 1.057 quarts
Gallons & m^3	1 gallon = 0.003785 m^3	1 m^3 = 264.2 gallons
Ounces & Grams	1 ounce = 28.35 grams	1 gram = 0.0353 ounces
Pounds & Kilograms	1 pound = 0.4536 kilograms	1 kilogram = 2.205 pounds
Tons & Pounds		1 metric ton = 2,205 pounds

Time Conversions
1 year = 52 weeks
1 week = 7 days
1 day = 24 hours
1 hour = 60 minutes
1 minute = 60 seconds

Volume and Cubic Measurements
1 quart = 2 pints = 57.5 in^3
4 quarts = 1 gallon = 231 in^3

ACKNOWLEDGMENT

Many thanks to Dr. Rebecca Ghent and Sarah Lebeau for the production of this laboratory exercise.

TOPOGRAPHY AND TOPOGRAPHIC MAPS

INTRODUCTION

The Earth's surface is characterized by hills and valleys, plains and plateaus, riverbeds, mountain ranges, and many other geological features. Topography, aka "relief," describes differences in elevation on Earth's surface. Examining where the high and low ground are located throughout a region has many practical applications ranging from military strategic planning to locating acceptable places for water wells and landfills. As such, maps depicting Earth's relief is critical for multiple reasons. Topographic maps are one method by which the three-dimensional topography of Earth's exterior can be represented on a two-dimensional (i.e., flat) surface.

OBJECTIVE

The objective of this lab is twofold: 1) introduce you to the concept of topography; 2) provide instruction on how to read and use topographic maps.

THE BEGINNING - TOPOGRAPHY AND TOPOGRAPHIC MAPS

I. **Latitude**: Global position relative to the equator (Figure 2.1).

II. **Longitude**: Global position relative to the prime meridian (Figure 2.1).

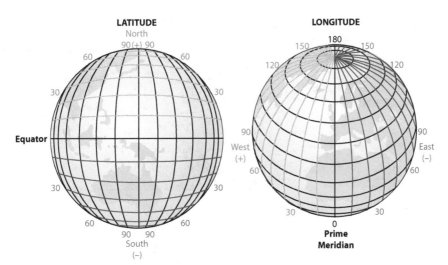

Figure 2.1 – Latitude and Longitude

11

III. **Datum**: A reference plane, or point of known location or value. For topographic maps, sea level is assigned an elevation of zero. All elevations are reported relative to mean sea level (msl) or zero.

IV. **Map Scales**: A way of relating distances measured on the map to distances in the real world.

    a. Fractional scale

       i.   Distance of 1 unit on the map represents a distance of "X" units on the Earth's surface.

       ii. For example, 1/24,000 or 1:24,000 indicates that one unit on the map is equivalent to 24,000 of the same unit in the real world.

       iii. Can be any unit (inches, feet, meters, km, etc.), so long as the unit is the same for both sides of the fraction.

    b. Graphic/Bar scale (Figure 2.2)

       i.   Line symbol subdivided into several parts representing distances on the Earth's surface.

       ii. Caution: Make sure you know where the "zero" is located—it may not be at the far left end of the scale.

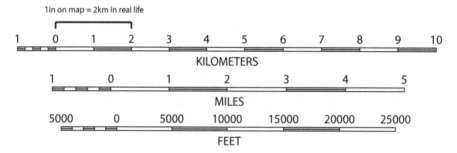

Figure 2.2 – Graphic/Bar Scale

V. **Contour lines** (Figure 2.3a and 2.3b): Lines drawn on a topography map that connect points of equal elevation.

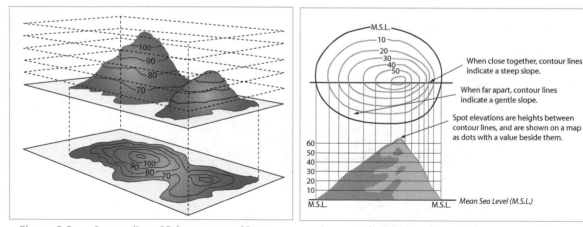

Figure 2.3a – Contour lines, 3D features on a 2D map.   **Figure 2.3b** – Contour lines, 3D features on a 2D map.

VI. **Contour interval** (Figure 2.4): Change in elevation between any two adjacent contour lines on the map. The contour interval is typically provided at the bottom of the map (see Figure 2.10). This interval is a fixed number for each map, but may differ from map to map. Map elevations are provided on *index* contours—typically every 5th contour line. For example, in Figure 2.4, index contour lines mark elevations of 0, 500, 1000, 1500, 2000, 2500, and 3000 meters. Elevations of all unmarked contours in between index contours can be deduced using the contour interval.

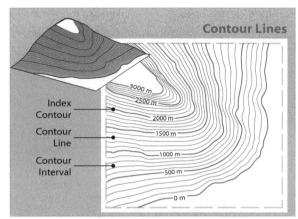

Figure 2.4 – Contour Intervals: Reading between the Lines.

- Every point on a contour line is an area of equal elevation. As a result of this, every contour line must eventually close on itself to form an irregular circle. Contour lines on the edge of a map do not appear to close on themselves because they run off the map. This being the case, if the map you are working with and the map representing the adjacent region were sitting side by side, you could follow the contour throughout its complete, enclosed path (refer to Figure 2.5).

- Contour lines can never cross one another. Each line represents a separate elevation, and you can't have two different elevations at the same point. The only time two contour lines may merge is if there is a vertical cliff (refer to Figure 2.5).

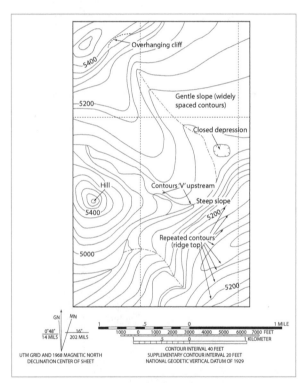

Figure 2.5 – Rules for Contour Lines

- Moving from one contour line to another always indicates a change in elevation. To determine if it's an uphill or downhill change, you must look at the index contours on either side (refer to Figure 2.5).

- Contour lines crossing a stream valley will form a "V" shape pointing in the uphill (and upstream) direction (refer to Figure 2.5).

- On a hill with a consistent slope, there are always four contours between two adjacent index contour lines. If there are more than four intermediate contours, it means that there has been a change of slope and one or more contour lines have been duplicated. This is most common when going over the top of a hill or across a valley (refer to Figure 2.5 and 2.7).

- The closer contour lines are to one another, the steeper the slope is in the real world. By the same logic, the farther away the contours are from one another, the milder the slope (see Figure 2.6). Evenly spaced contours indicate a constant slope, while unequally distributed ones demonstrate a change of slope.

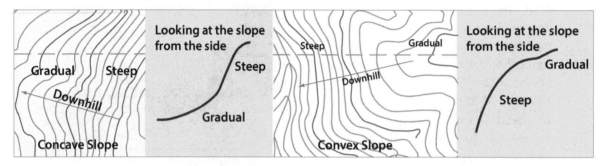

Figure 2.6 – Contour Lines and Slopes

- A series of closed contours (the contours make a circle) represents a hill. If the closed contours are hachured, it indicates a closed depression (refer to Figure 2.5). Note that the contour line repeats around hills, valleys, and when entering a depression (see Figure 2.7).

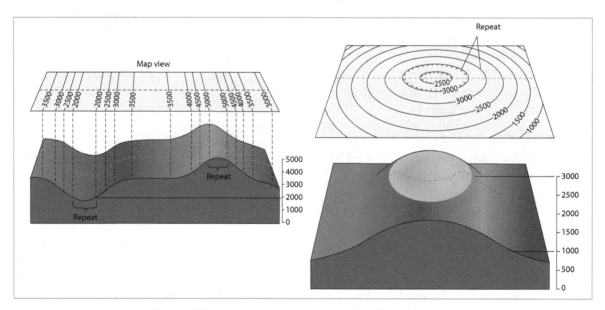

Figure 2.7 – Contour Lines representing Hills, Valleys, and Depressions

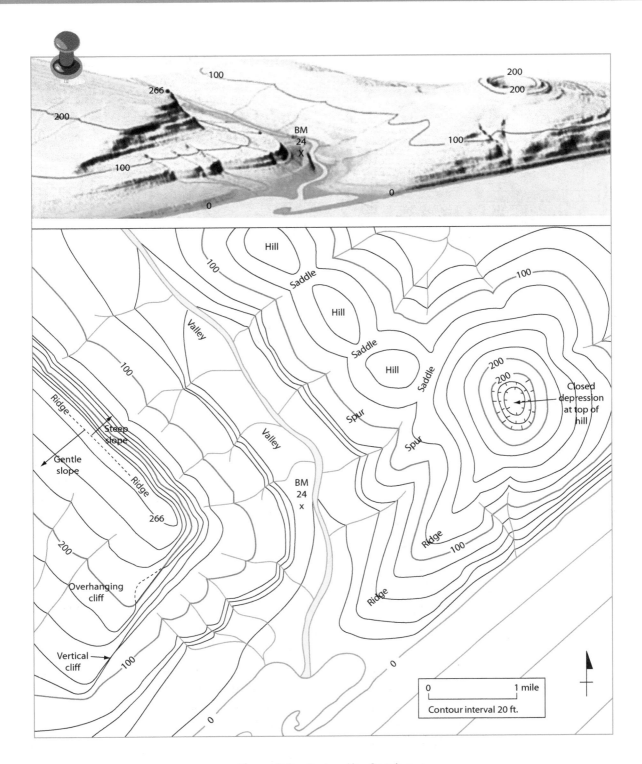

Figure 2.8 – Contour Line Overview

VII. **Relief:** The difference in elevation between two points on a map.

VIII. **Gradient** = $\dfrac{\text{Change in elevation or relief (measured using contour lines)}}{\text{Change in horizontal distance (measured using map scale)}} = \dfrac{\text{rise}}{\text{run}}$

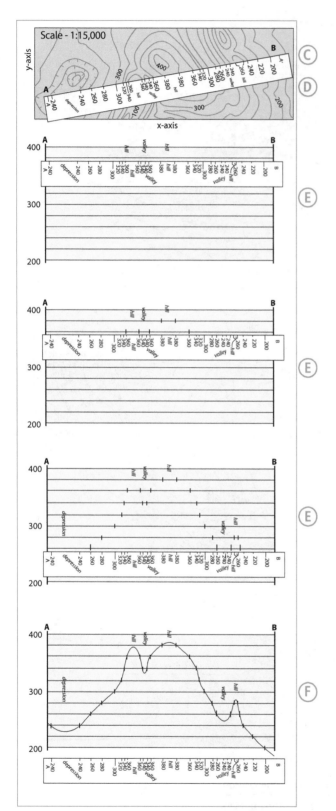

Figure 2.9 – How to draw a topographic profile

IX. **Vertical Profile or Vertical Section**: A representation of a vertical slice through Earth's surface. This is done across a traverse line between two designated points on the map (ex: A–B). See Figure 2.9 for a schematic representation of constructing a vertical profile.

CREATING A VERTICAL PROFILE

Ⓐ
Locate the traverse line, which will serve as the x-axis.

Ⓑ
Using the elevations on the topographic map, decide on an appropriate scale for the Y-axis of the profile graph, and label it accordingly. This can be done by putting the lowest and highest elevations on the graph, then marking even increments in between.

Ⓒ
Align a straight piece of paper along the traverse line.

Ⓓ
For every place the traverse line crosses a contour line, place a tick mark at the top of the paper and label the tick mark with the elevation the contour line represents.

Ⓔ
Once all the intersections have been marked, align the paper with the x-axis of the profile graph. Next, drop down vertically from each tick mark and put a point on the graph at the appropriate elevation represented by each tick mark.

Ⓕ
Connect the dots.

Ⓖ
Calculate the vertical exaggeration. See Roman Numeral X for an explanation of this procedure.

X. **Vertical Exaggeration**: A representation of the amount of exaggeration along the vertical (Y) axis on a profile compared to the horizontal (X) axis. To determine the VE, simply divide the vertical scale by the horizontal scale.

$$\frac{\text{Vertical scale}}{\text{Horizontal scale}} = \text{Vertical exaggeration}$$

For example: If the vertical scale is 1:3500 and the horizontal scale is 1:35,000, the vertical exaggeration is:

$$\frac{1/3500}{1/35000} = \frac{35000}{3500} = 10x$$

This means that the profile plot is distorted … as it is stretched 10x more in the vertical direction than in the horizontal.

HINT: Watch YOUR UNITS!!!

*See Figure 2.10 on the following page for a standardized layout of a topographic map.

Figure 2.10 – Topographic map

Western Longitude
109° 00' 00"

Northern Latitude
38° 15'

Northern Adjacent Map

Eastern Longitude
108° 52' 30"

Northern Latitude
38° 15'

Southern Latitude
38° 07' 30"

Western Longitude
109° 00'

Fractional Scale

Southern Adjacent Map

Bar Scale

Contour Interval

Datum

Southern Latitude
38° 07' 30"

Eastern Longitude
108° 52' 30"

Adjacent Maps

1	2	3
4	5
6	7	8

1. Ray Mesa
2. Paradox
3. Davis Mesa
4. Lisbon Gap

5. Bull Canyon
6. Summit Point
7. Horse Range Mesa
8. Hamm Canyon

PART A: LEARNING TO USE TOPOGRAPHIC MAPS

Problems

1.

a. In Figure 2.11, fill in the boxes on the contour lines with the correct elevation in feet above sea level. (*Note:* Contour interval is 20 ft.)

b. Using a red pencil, color in the area on the figure that represents the lowest elevation.

c. Label the closed depression by putting a ★ symbol in the middle.

Figure 2.11 – Learning to use contour lines

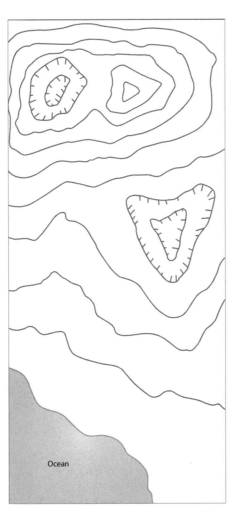

2. Using a contour interval of 10m, label every line on Figure 2.12 with its exact elevation above sea level. Note: Before attempting this exercise, re-examine figure 2.7 for a memory refresher on labeling closed depressions.

Figure 2.12 – Learning to use contour lines

3. Use Figure 2.13 to answer the
    questions that follow:

    a. The contour lines on this map
        are labeled in feet. What is the
        index contour interval on this
        map? _____ ft

    b. What is the contour interval on
        this map? _____ ft

    c. What is the regional relief, i.e.,
        total relief of the land
        represented on this map?
        _____ ft

    d. What is the gradient from A to
        B? _____ ft/mile
        Show your work.

Figure 2.13 – Working with Topo Maps

e. Two potential routes to the top of a hill are marked on Figure 2.13. One is in blue and one is
    in green. Which of these two routes is more strenuous? The blue or the green? Why?

    _____

    _____

    _____

f. How could you find the areas on this map that have a gradient of 20 ft/mile or greater?
    (*Hint:* Think of the contour interval and how many contour lines of map elevation must
    occur along one mile of map distance.) _____

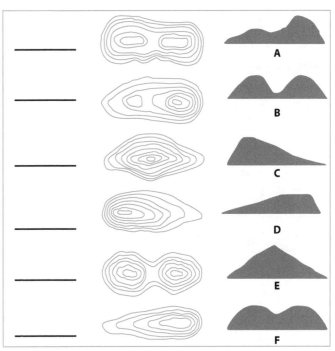

4. Match each of the contour sketches in Figure 2.14 with the land features they represent by putting the letter of the appropriate land feature on the line provided.

Figure 2.14 – Land features and contour lines

5. Draw a vertical profile of the A–B traverse line on Figure 2.15 *and* calculate the vertical exaggeration. Note: all elevations are in feet.

    Show your work for the Vertical Exaggeration calculation:

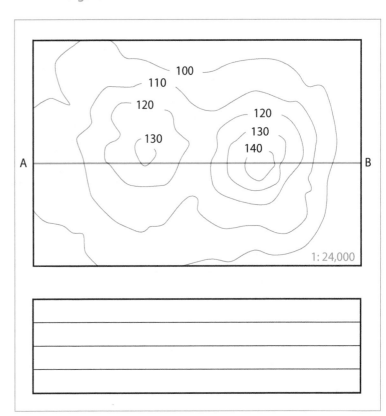

VE = _____

Figure 2.15 – Drawing Topographic Profiles

PART B: WORKING WITH TOPOGRAPHIC MAPS

Use the Washington, D.C. quadrangle to answer the following questions.

6. If you wanted to see the topography of the area to the northwest of this map, which map would you need to view? _____

7. What is the contour interval? _____ ft.

8. If the contour interval were twice what it actually is, would there be a **greater** or **lesser** number of contour lines? (Circle one)

9. Index contour lines appear every _____ feet on this map.

10. What is the general flow direction of the Potomac River and Rock Creek? (Circle the correct answer.) S/SE, S/SW, N/NE, N/NW

11. Express the scale of this map in four ways:
    a. The fractional scale of this map is _____.

    b. One inch on the map = exactly _____ inches in real life.

    c. One inch on the map = exactly _____ miles in real life.

    d. One inch on the map = exactly _____ km in real life.

12. Name the parallels of latitude that serve as the northern and southern boundaries of the map
    North: _____   South: _____

13. Name the meridians of longitude that serve as the western and eastern boundaries of the map
    West: _____   East: _____

14. How many degrees of both latitude and longitude are covered on this map?
    Lat: _____   Long: _____

Show your work.

Use the topographic map of Front Royal, Virginia, to answer the following questions. *Show your work.*

15. You're riding your bike in Shenandoah National Park on Skyline Drive. What gradient (in feet per mile) would you experience between the North Entrance, (near Front Royal), to the first stream crossing (see Figure 2.16 for a visual depiction of the area in question)? *Show your work.* _____ ft/mile

16. Notice there is a series of streams surrounding the words "Shenandoah National Park" at the bottom, center of the map. See Figure 2.17a for a visual depiction of the area in question. In which direction are these streams flowing?

    _____

    How do you know?

    _____

    _____

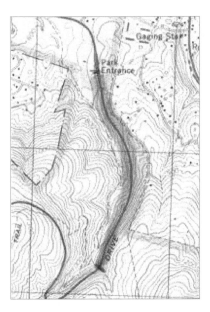

Figure 2.16 – Gradient

17.

Pluto and Goofy decide they want to go out and get some exercise and jointly agree that a moderate bike ride would be just the thing they needed. After looking at Front Royal topo map they concluded that a ride along the A–A′ line looked very promising (see Figure 2.17a).

Figure 2.17a – Profile Line for Prospective Path

Just to make sure they weren't getting in over their heads, they quickly plotted out a profile of the transect so they could get an idea of just how steep the terrain is. Figure 2.17b is the profile they drew up using a vertical scale of 1 inch = 200 ft.

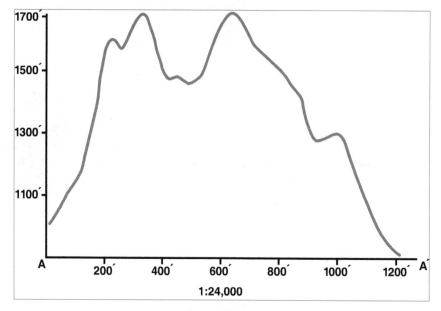

Figure 2.17b – Profile with vertical exaggeration

a. Based on their profile, does this look like a ride two amateur bike riders should be taking? Yes or No? Explain.

_____

b. Calculate the vertical exaggeration of this profile plot. *Show work below.*

$VE =$ _____

Now compare their profile to a profile drawn to scale (1 inch = 2,000 ft). See Figure 2.18.

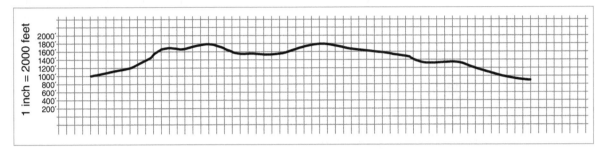

Figure 2.18 – Profile drawn to scale

1 inch = 2,000 ft

2,000 ft = 24,000 inches

Therefore, Vertical Scale is 1 inch = 24,000 inches

1/24,000 (Vertical Scale)
1/24,000 (Horizontal Scale)

VE = 1:1 ratio = no vertical exaggeration

c. What is your opinion now? _____

_____

18. If the South Fork of the Shenandoah River near Front Royal flooded 50 feet above normal due west of the Randolph Macon Military Academy, name four sites that would be flooded. _____, _____, _____, _____.

19. Assume that the elevation of the water surface in the river reaches ten feet below the adjacent land contour. Which specific sites pose a potential environmental threat in the event a flood occurs? _____ and _____.

Part C: Tying it all Together

On the center of each table is a plastic tub covered with a lid into which holes have been drilled at equal intervals. Inside each tub is a replicate 3D landscape full of mountains, valleys, stream beds, and other geographical features. You are going to create a topographic map of this 3D landscape by following the procedure below:

1. Examine figure 2.19. Figure 2.19 is a diagram that represents the lid of your tub. Each of the holes on the diagram corresponds to a hole in an identical location on the tub's lid. Starting at one of the corner holes on the tub's lid, use the stick that has been supplied for you to measure the vertical distance (elevation) between the top of the landform and the rim of the hole. You do this by inserting the stick into the hole and lowering it until it can go no further. Read the measurement from the ruler on the stick, then record the elevation (in cm) by the appropriate hole on figure 2.19. Repeat this procedure for all the holes in the lid.

2. Using a 2cm contour interval, connect all the holes of equal elevation. Don't forget to label your index contour lines. Your TA will also be around to help with this, as it can be a little confusing for beginners. If after you've given it a fair shot yourself and are still having problems, call over your TA and he/she will be happy to assist you.

Once the lines are all connected up, give yourself a sound pat on the back – as you have just created your first topographic map.

3. Using figure 2.20, create a vertical profile of the A – A' line on your self-made topography map.

4. Calculate the vertical exaggeration in the space provided below.

Vertical Exaggeration – _____x

5. NOW ... TAKE THE TOP OFF OF THE TUB AND SEE IF YOUR TOPOGRAPHIC MAP AND PROFILE ACCURATELY DEPICT THE LAND FORMATIONS!!!

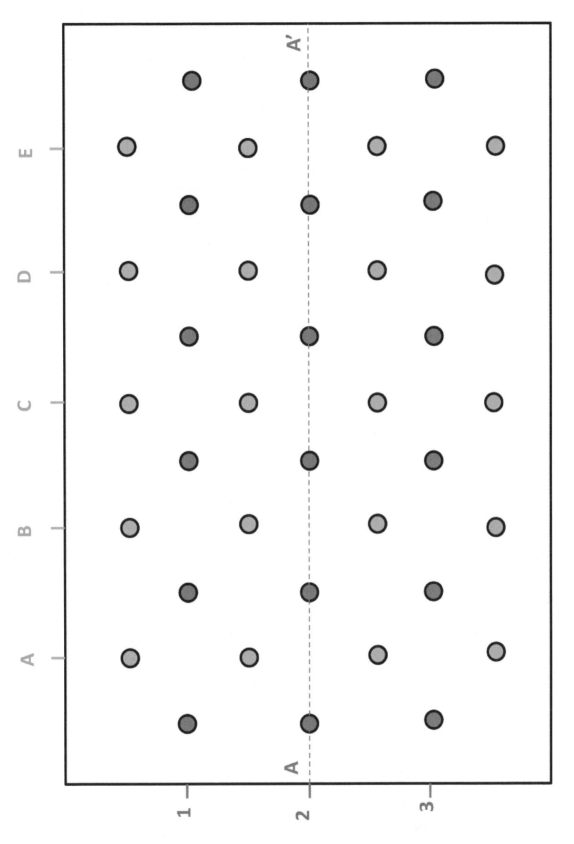

4cm = 6cm

Figure 2.19

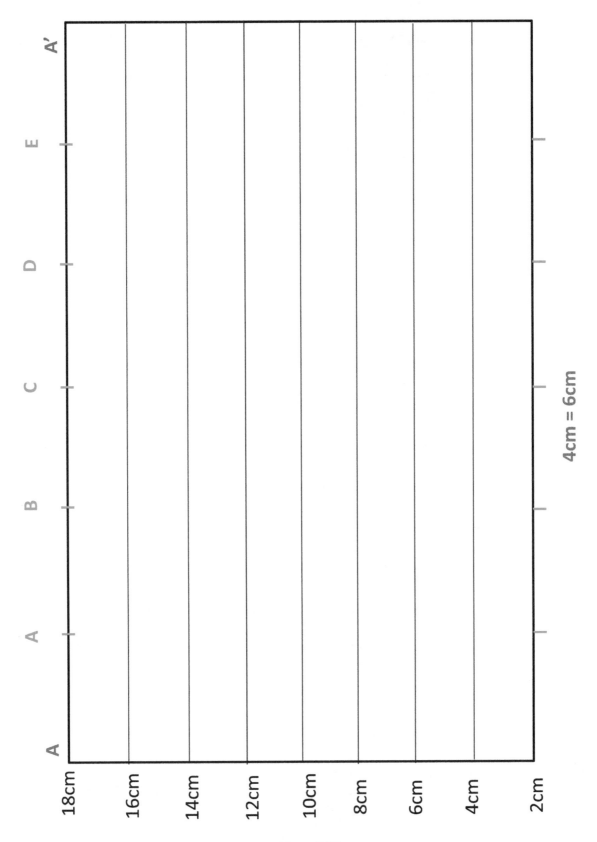

Figure 2.20

MINERAL IDENTIFICATION

INTRODUCTION

Given that the Greek root *geo* means "world" and *ology* means "study of," the word geology, translates to "study of **EARTH**." It is therefore the job of a geologist to investigate how nutrients, resources, and energy cycle through Earth's multiple subsystems (lithosphere, hydrosphere, and atmosphere), and ultimately contribute to the sustenance of a thriving biosphere. While there are numerous mechanisms by which these necessities become available for biological consumption, the most fundamental—and in many ways the most important—is what we call a mineral.

Minerals are essentially the building blocks of the Earth. It is minerals that compose the rocks that make up our planet, serve as the main ingredient in soil and are therefore the supplier of nutrients to the biosphere, and provide countless resource materials we use on a daily basis. As such, whether grown or mined, rest assured its origins were in what we call a mineral.

Minerals differ from other forms of matter in that they must satisfy the following five criteria:

1. Be capable of being produced by nature
2. Retain a solid form at standard surface temperatures and pressures
3. Be inorganic (lacking a C - H bond)
4. Have a crystalline structure—i.e. an orderly internal arrangement of atoms that repeats at regular intervals
5. Have a fixed chemical formula

To date, there are over 4000 different minerals, which are categorized into 8 families (see Tables 3.A and 3.B) but a working knowledge of 10–20 of them will allow you to identify most rocks by their composition and therefore gain an understanding of how Earth is built.

OBJECTIVE

The objective of this lab is three-fold:

1. Familiarize you with the diagnostic properties of minerals
2. Acquaint you with several of the most common rock forming minerals
3. Introduce you to some of our most prominent resource minerals.

Mineral/Formula	Example
Olivine group $(Mg, Fe)_2SiO_4$	Olivine
Pyroxene group (Augite) $(Ca, Mg, Fe, Ti, Al)_2(Si, Al)_2O_6$	Augite
Amphibole group (Hornblende) $Ca_2(Fe, Mg)_5Si_8O_{22}(OH)_2$	Hornblend
Micas — Biotite $K(Mg, Fe)_2AlSi_3O_{10}(OH)_2$	Biolite
Micas — Muscovite $KAl_2(AlSi_3O_{10})(OH)_2$	Muscovite
Feldspars — Potassium feldspar (Orthoclase) $KAlSi_3O_8$	
Feldspars — Plagioclase feldspar $(Ca, Na)AlSi_3O_8$	
Quartz SiO_2	Potassium Feldspar and Quartz

Table 3.A – Silicate Family

Mineral Family	Example
Oxides (O)	Hematite (Fe_2O_3) Magnetite (Fe_3O_4)
Sulfides (S)	Pyrite (FeS_2) Galena (PbS)
Sulfates (SO_4)	Gypsum $(CaSO_4$–$2H_2O)$
Phosphates (PO_4)	Apatite $Ca_5(PO_4)_3(Fl, Cl, OH)$
Carbonates (CO_3)	Calcite $(CaCO_3)$
Halides $(F^-, Cl^-, Br^-, I^-, or At^-)$	Halite $(NaCl)$ Fluorite (CaF_2)
Native Elements	Gold (Au), Silver (Ag), Platinum (Pt)

Table 3.B – Non-Silicate Families

PART A: DIAGNOSTIC PROPERTIES OF MINERALS

How do we tell one mineral from another? Well, just like you have a certain hair color, eye color, height, weight, and body shape that distinguish you from another person, each type of mineral has a unique set of physical properties that distinguishes it from all other types of minerals.

I. Why do different types of minerals exhibit different properties?

A mineral's physical properties (ex: hardness, shape, color, etc.) are determined by A) the mineral's crystal structure—i.e., the arrangement and bonding of the atoms in the mineral; and B) the mineral's composition—i.e., the types of atoms in the crystal structure. If the crystal structure and/or composition differ from one sample to another, then the samples are going to have distinctly different physical properties. These similarities and differences in physical properties, resulting from variance in crystal structure and composition, are how we group like samples together as one type of mineral and differentiate them from other types of minerals.

II. What properties help us determine the identity of a mineral?

Properties Due to Crystal Structure

Crystal Form (Habit): Refers to the general shape or character of a crystal or cluster of crystals that have grown in nature unimpeded. Crystal forms are classified by such terms as fibrous, needlelike, prismatic, cubic, and platy to name a few. Crystal form is a direct reflection of the arrangement of atoms within the crystal structure, and therefore will vary between minerals. Crystal form can be an efficient diagnostic tool and is often useful when differentiating between minerals that otherwise appear nearly identical. See Figure 3.1 for select examples of crystal habit.

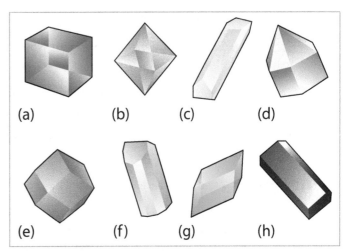

Figure 3.1 – Selected crystal forms.

Cleavage and Fracture: Describes the method by which a mineral breaks. Many minerals break in an identifiable, orderly manner along flat surfaces. These surfaces are called "cleavage planes" and result from weak chemical bonding between parallel layers of atoms in the crystal structure. Because different crystal structures have weak bonds located in different places, minerals with dissimilar crystal structures will have diverse cleavage patterns. Some structures have no weak bonds at all, others have only one set of weak bonds, and still others have multiple sets of weak bonds. Each different set of parallel cleavage planes has an orientation relative to the crystal structure and is referred to as a "cleavage direction." See Figure 3.2 for a visual representation of select cleavage examples.

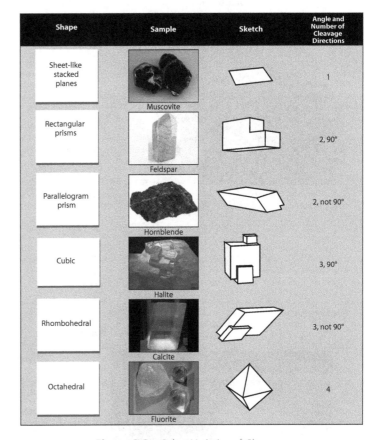

Figure 3.2 – Select Varieties of Cleavage

• Cleavage is described as either:

Excellent – Easily identifiable; minerals that when broken produce pieces shaped as cubes, rhombo-hedrons, columns, pyramids, or planes.

Good – Minerals that when broken retain a distinctive shape with flat sides, but have distortion around parts of the mineral (ex: rounded edges).

Poor/Absent – Distinguishing between minerals with poor cleavage and no cleavage is many times not possible, as poor cleavage surfaces are often inconspicuous.

★Note that when a mineral cleaves, it produces numerous broken pieces of similar or identical shape.

Minerals that have bonds of comparable strength within their crystal structures will fracture, as opposed to cleave, when stressed. This results in the mineral randomly shattering instead of breaking along defined boundaries, so no two broken pieces will be the same shape.

Fracture is described as either conchoidal or non-conhoidal (see Figures 3.3 and 3.4):

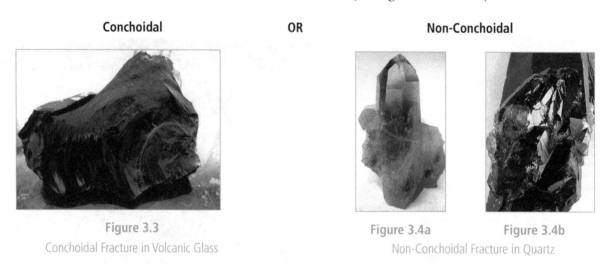

Conchoidal	OR	Non-Conchoidal

Figure 3.3
Conchoidal Fracture in Volcanic Glass

Figure 3.4a   **Figure 3.4b**
Non-Conchoidal Fracture in Quartz

★Note that the absence of cleavage can be equally as diagnostic as the presence of cleavage.

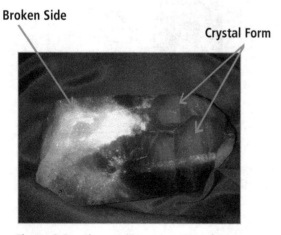

Broken Side

Crystal Form

Figure 3.5 – Cleavage/Fracture vs. Crystal Form

Cleavage can often be confused with crystal form, but that can easily be rectified by looking at a broken side. If the mineral has been broken, then excellent/good cleavage (or fracture) will be visible. If however, the mineral is completely intact as nature produced it, then what you are seeing is crystal form (Figure 3.5). In addition, cleavage planes can be repeated like steps, whereas a crystal face is one solid, smooth surface.

Mineral Hardness: A measure of the resistance of a mineral to abrasion or scratching. A mineral's hardness is determined by: A) the strength of the bonds throughout the crystal structure; and B) the configuration of the atoms within the crystal structure. Minerals made up of atoms that are held together primarily by the weak van der Waals force constitute our softest minerals, as it takes very little energy to terminate the attraction between the atoms and break up the structure (ex: talc and graphite). Minerals comprised primarily of strong covalent bonds (ex: quartz and the diamond) are the hardest minerals on Earth, and those made up primarily of ionic bonds (halite and calcite) fall in between, albeit closer to the softer end. In addition, the more densely packed the atoms, the stronger the bonding and the harder the mineral.

Relative mineral hardness is categorized using Mohs Hardness Scale (Fig. 3.6), which assigns each mineral a number between 1 and 10—1 being the softest and 10 being the hardest. Absolute mineral hardness can be read from the Y-axis and divulges exactly how much harder each mineral is than another.

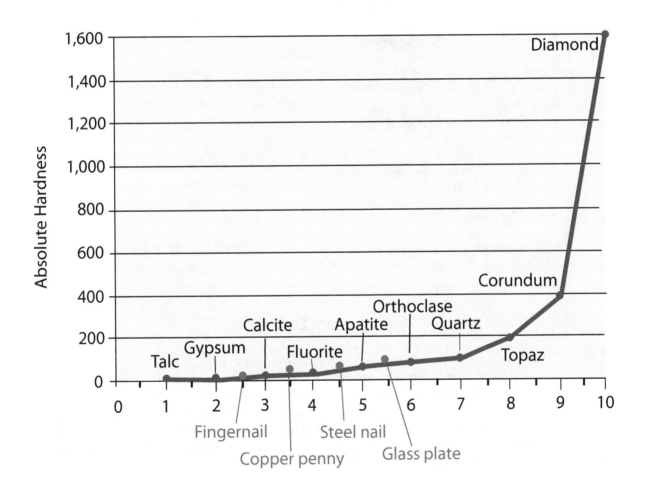

Figure 3.6 – Relative and Absolute Hardness

How significantly does the type of crystal structure and bonding affect the properties of a mineral?

Let's look at an example where two minerals have the same chemical composition, but very different structures—the diamond (C) and graphite (C). See Figure 3.7.

Diamond (C)   vs.   Graphite (C)

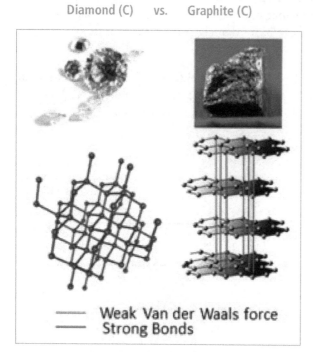

Weak Van der Waals force
Strong Bonds

Figure 3.7 – Diamond vs. Graphite

Diamond (C)

- Made exclusively of covalent bonds
- 3D crystal structure
- Dense → Due to compact packing of atoms
- Hardest material on Earth
- Electrical Insulator

Graphite (C)

- Has van der Waals force between sheets
- Sheet crystal structure
- Not dense → Due to spacing between the atoms
- One of the softest materials on Earth
- Electrical Conductor

• So … the difference between what you buy in Staples for $3.99 a pack and the "forever" gem that costs upward of $2000 is nothing more than the orientation of the atoms and the strength of the bonds.

Properties Due to Composition

Color – Color results from the way a mineral interacts with light. Sunlight contains the whole spectrum of color and each color has its own unique wavelength. The mineral absorbs most of the colors' wavelengths, and the color you see when looking at it represents the wavelength the mineral did not absorb (i.e., the one it reflected). Certain minerals always have the same color, but many show a range of colors due to minute impurities that slide into the crystal structure. As such, in many cases, color is not a diagnostic property of a mineral, as one mineral may have multiple colors and multiple minerals may have the same color (See Figures 3.8 and 3.9).

Figure 3.8 – Select Color Varieties of Quartz
Select quartz family members including amethyst, citrine, rose, colorless rock crystal, chalcedony, tiger eye, onyx, agate and aventurine.

Hematite

Magnetite

Graphite

Figure 3.9 – Three Different Minerals

Figure 3.10– Sulfur Streak

Streak – Color of the fine powder of a mineral, obtained by rubbing the mineral across a piece of unglazed porcelain (see Figure 3.10). Pulverizing the mineral into powder form, will result in all the minerals of the same type having a similar color streak, regardless of the external color. Note: The streak plate has a hardness on Mohs Hardness Scale of ~6.5 –7. As such, minerals harder than 6.5 cannot be tested for streak. Given what you now know about mineral hardness, consider why this is the case.

Figure 3.11 – Magnetite Attracts Iron Nails. Photograph from 1921

Magnetism – There is a limited selection of minerals that demonstrate magnetic properties. The presence of iron in the mineral's chemical composition is single-handedly responsible for the unique behavior of any magnetic mineral. The most well-known magnetic mineral is Magnetite, hence the descriptive name (see Figure 3.11).

Reaction to hydrochloric acid – When a member of the carbonate family (group of minerals whose chemical formula includes a CO_3 group) comes in contact with dilute hydrochloric acid, the acidic aqueous solution has enough strength to break the covalent bonds holding the carbonate group together. This reaction produces CO_2 as a byproduct, which manifests itself in the form of erupting gas bubbles—aka "fizzies,"—see Figure 3.12. While all carbonate minerals fizz to some degree upon contact with HCl, the carbonate mineral you will be working with is known as calcite ($CaCO_3$). Note: If one drop of dilute HCl solution does not make the mineral fizz, then 10 drops will not make it fizz … so please do not bathe the minerals in HCl.

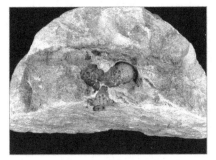

Figure 3.12 – Calcite-rich rock reacts with HCl

Specific Gravity – Measures the density (weight per unit volume) of a mineral. This can be a useful diagnostic tool, as samples of two different minerals that are the same size and of similar appearance may have significantly different weights.

Properties Due to BOTH Crystal Structure and Composition

Luster – Describes the way a mineral reflects light. The most basic division of luster is metallic versus nonmetallic (see Figures 3.13 and 3.14).

Metallic:
Silvery, gold, or brass sheen displayed by metals

Nonmetallic:
Any luster unlike that of a metal.

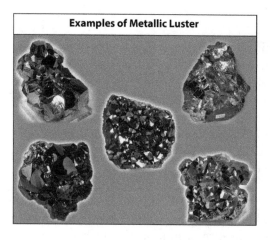

Figure 3.13 – Metallic Luster

Note: While every mineral has a characteristic luster, minerals of the same type may exhibit different lusters. In addition, there is no scientific method available to determine luster. This being the case, determining the luster of a particular specimen is subjective; two people looking at the same specimen may observe two different lusters. For these reasons, we are going to restrict our analysis of luster to Metallic vs. Nonmetallic only.

Examples of Non Metallic Luster	
Nonmetallic: Earthy/Dull • Non-reflective, like concrete or dirt • Develops when minerals are composed of coarse grains that scatter light in all directions.	
Nonmetallic: Greasy • Resembles fat or grease • Results from a significant amount of microscopic impurities included in the structure.	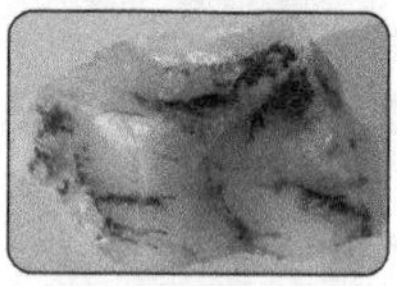
Nonmetallic: Pearly • Thin transparent sheets stacked one top of the other give the mineral a pearl like appearance	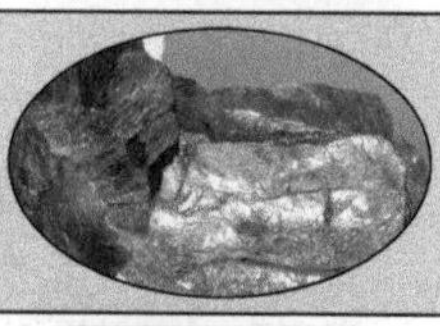
Nonmetallic: Resinous • Has the appearance of resin or smooth plastic	
Nonmetallic: Vitreous • Like glass or a glossy photo • Typically found in transparent and translucent minerals	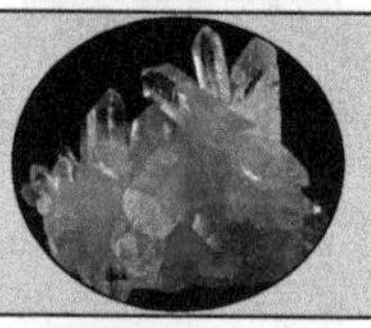
Nonmetallic: Silky • Very fine fibers align parallel giving the appearance of silky fabric	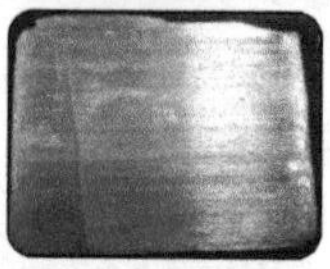
Nonmetallic: Waxy • Has a candly-like texture	

Figure 3.14 – Non Metallic Luster

Diaphaneity – A measure of the amount of light that passes through a mineral (see Figure 3.15). This is determined by the amount of light being absorbed by the mineral—which is, in turn, a function of its composition and crystal structure.

• Diaphaneity is described as either:

Transparent – No light is absorbed, permitting all the light to pass through the mineral. Completely transparent minerals do not have color.

Translucent – Allows light to pass through, but scatters and diffuses it, creating only a partial transparency.

Opaque – All light is either absorbed or reflected, meaning no light passes through the mineral.

Transparent

Translucent

Opaque

Figure 3.15 – Comparison of Diaphaneity

Exercise: Using Diagnostic Properties

While many different minerals may appear similar or even identical at first glance, a closer examination using the properties described above will demonstrate they are in fact very different materials. On the flip side, minerals that look nothing alike, but exhibit the same properties will be shown to be the same substance. The box on the table in front of you, labeled "mixed minerals," contains 14 mineral samples. The samples in this box are made up of 6 different minerals, meaning there are several samples of each mineral. Your job is to divide up the 14 samples into 6 groups, each comprising the samples made of that mineral only. In the chart below, record the letters of the samples that belong to each group and write down what properties made you group those samples together and differentiate them from the others. The number of samples per mineral has been supplied for you. Once you have completed this task, your TA will go over the answers with you, before proceeding to the next section.

# of Samples in Group	Letter of Samples that belong to this group	Properties that unite these samples and distinguish them from the others
4		
3		
2		
2		
2		
1		

PART B: MINERAL IDENTIFICATION

This exercise is meant to help you learn how to identify minerals using the properties described in Part A. In the container before you are multiple unidentified mineral samples. Your jobs is to identify them using the procedure below. First, list the properties of each of the mineral samples in the accompanying Mineral Identification Charts provided on the next pages. Next, use the flow charts at the end of this lab to help you identify each mineral. The flow charts are designed to allow you to use the most basic properties, like cleavage, to first rule out impossible samples and then to further narrow down the mineral's identity using other less-obvious properties, like hardness, streak, specific gravity, etc. Keep in mind that not all properties of every mineral are equally useful. Generally you will be able to identify a mineral by using only two or three primary properties. These will differ for each mineral. For instance, although galena has a hardness of 2.5, this is not generally a primary property, as many other minerals also have this same hardness. You will likely be able to identify galena by the fact that it has a very high specific gravity and has cubic cleavage (three cleavage planes), thus the density and cleavage (cubic) will be the primary properties for galena. In addition, some minerals depict one specific property (such as magnetism) that can lead you to an immediate identification. Determine which properties are most valuable in helping you identify each mineral and enter these properties in the primary properties column of your chart.

Name _____

MINERAL IDENTIFICATION CHART

Specimen Number	Color	Streak	Hardness	Cleavage (good or absent)	Primary Properties	Name
1						
2						
3						
4						
5						
6						
7						
8						
9						
10						

Name _____

MINERAL IDENTIFICATION CHART

Specimen Number	Color	Streak	Hardness	Cleavage (good or absent)	Primary Properties	Name
11						
12						
13						
14						
15						
16						
17						
18						
19						
20						

Now that you have completed the mineral charts, answer the questions below.

I.   Using the definition of a mineral, determine whether each of the following is or is not a mineral; if the substance is not a mineral, explain why.

   a. Wood

   b. Halite (NaCl, also known as salt)

   c. Mercury

II. Graphite (hardness of 1) and diamond (hardness of 10) are both made of carbon atoms. Thinking about the definition of a mineral, why might their respective hardnesses be so different?

   _____

   _____

   _____

   _____

III. Olivine, feldspar, mica (muscovite and biotite), amphiboles, pyroxenes, and quartz are among the most common rock-forming minerals. To which mineral class(es) do these minerals belong?

   _____

Mineral Name	Cleavages	Color	Formula	Habit	Hardness	Helpful Diagnostic Properties	Luster	Opacity	Streak	Uses
Augite (Pyroxene)	2 directions of cleavage at nearly right angles	Black or greenish-black	Ca, Mg, Fe, Al silicate	Forms columns Generally occurs as anhedral to subhedral crystals in matrix	5–6	Cleavage	Vitreous or resinous	Opaque	Greenish-gray	N/A
Bauxite	Cleavage poor	Pale to dark reddish-brown	Hydrous Al Oxide	Pisolitic to massive	1–3	Covered in "bulls-eyes"	Earthy	Opaque	N/A	Manufacture of aluminum alloys
Biotite Mica	Excellent sheet cleavage	Black, dark brown	$K(Mg,Fe)_3 (Al,Fe) Si_3 O_{10} (OH,F)_2$	Platy Sheets	2.5–3	Cleavage + color	Sub-metallic, vitreous	Transparent to opaque	Gray-Brown	Used for insulating in electrical equipment, in lubricants, wall finishes, artificial stone.
Calcite	Excellent rhombohedral cleavage	Colorless white, gray, yellowish, or blue	$CaCO_3$	Prisms, rhombohedrons, scalenohedrons	3	Fizzes with HCl	Glassy or milky	Opaque to transparent	White	Used as fertilizer, in cement manufacture, filler in paper, rubber, etc., and in plaster.
Galena	Good cubic cleavage	Lead-gray	PbS	Crystals usually cubic	2.5	Extremely dense	Bright metallic	Opaque	Lead-gray	Main ore of lead. Pipes, alloys, batteries, radiation shielding, pigments.
Graphite	1 good cleavage direction	Iron-black to steel-gray	C	Massive to platy	1–2	You can write with it (Has a greasy feel)	Metallic to dull	Opaque	Black	Manufacture of crucibles, paint, and foundry molds. Used as lubricant and electric furnace electrodes. Writing part of pencils.
Gypsum	Cleavage present	Colorless, white, gray, yellowish	$CaSO_4 - 2H_2O$	Sheets, fibrous, aggregates of small crystals	2	Scratch it with fingernail	Sub-vitreous, pearly on cleavage	Transparent	White	Used as fertilizer, in cement manufacture, as a filler in paper, rubber, etc., and as plaster.

Mineral Name	Cleavages	Color	Formula	Habit	Hardness	Helpful Diagnostic Properties	Luster	Opacity	Streak	Uses
Halite	Cubic	Colorless, white, yellow, orange, reddish, purple, blue	$NaCl$	Crystals cubic	2	Tastes salty	Vitreous	Transparent to translucent	White	Used in foods, road de-icing, and chemical manufacturing
Hematite	Poor or absent	Silvery-gray, black, or brick red	Fe_2O_3	Massive or thin tabular crystals	1.5–6	Streak	Earthy or metallic	Opaque	Red or red-brown	Principle ore of iron and as a gemstone.
Hornblende (amphibole)	2 directions at nearly 60° and 120°	Black or greenish-black	Ca, Na, Mg, Fe, OH, Al Silicate	Common crystals are prismatic	5–6	Cleavage	Vitreous	Opaque	Grayish-white	N/A
Magnetite	1 good cleavage direction (not always visible)	Iron-black, grayish-black	Fe_3O_4	Crystals octahedral and commonly granular (magnetic)	6	Magnetic	Metallic to dull	Opaque	Black	Major source of iron and therefore steel. Medicinal properties that are used to treat various medical conditions.
Muscovite Mica	Excellent sheet cleavage	Colorless, gray, green	$KAl_2(Si_3Al)O_{10}(OH,F)_2$	Platy Sheets	2.5–4	Cleavage + color	Vitreous to pearly or silky	Transparent to translucent	White/clear	Used for insulating in electrical equipment, in lubricants, wall finishes, artificial stone.
Native Copper	N/A	Copper-red, brown; often tarnished in a pistachio green	Cu	Crystals cubic, octahedral, dodecahedral, tetra-hexahedral; arborescent, wirelike, massive, powdery	2.5	Green tarnish	Metallic	Opaque	Pale red	Alloyed with various metals to produce bronzes and brasses. Used in electrical industries where high electrical and thermal conductivity is required.
Olivine	Present but poor	Olive green	$(Mg,Fe)_2SiO_4$	Usually occurs as rounded grains, in dense aggregates of grainy crystals, and as fractured masses	6.5–7	Usually tiny crystals that flake off like sand. Sometimes gem quality	Vitreous	Transparent to translucent	White/clear	The variety Peridot is a famous gem. Olivine is also used as a flux for making steel, and is an ore of magnesium.

Mineral Name	Cleavages	Color	Formula	Habit	Hardness	Helpful Diagnostic Properties	Luster	Opacity	Streak	Uses
Plagioclase Feldspar	2 directions of cleavage at nearly right angles	White, gray, blue	$(Na,Ca)(Si,Al)Si_3O_8$	Striated tabular crystals or blades	6–6.5	Striations	NM. Vitreous, often pearly on cleavage	Opaque	White/clear	Manufacture of porcelain, pottery and glass. Used for glazes in pottery and as mild abrasive.
Potassium (K) Feldspar	2 directions of cleavage at nearly right angles	Pink, peach, cream, turquoise	$KAlSi_3O_8$	Crystals short prismatic, blocky, or tabular; massive, cleavable to granular	6–6.5	Cleavage + color	NM. Vitreous, often pearly on cleavage	Opaque	White/clear	Manufacture of porcelain, pottery and glass. Used for glazes in pottery and as mild abrasive.
Pyrite	Cleavage poor	Brass-yellow	FeS_2	Crystals usually cubic	6	"Fool's Gold"	Metallic	Opaque	Goldish-brown	Iron and steel
Quartz	None Conchoidal fracture	Colorless, white, gray, yellow to brown to black, violet, pink	SiO_2	Crystals short to long prismatic, elongated along c-axis, hexagonal, horizontally striated, bent, distorted, skeletal	7	Scratches glass	NM. Vitreous, sometimes greasy or waxy	Transparent to translucent	White/clear	Used in building materials, glass-making, pottery, ferro-silicon alloys, as abrasives and in electronics.
Sulfur	Cleavage poor	Sulfur-yellow to yellowish-brown, yellowish-gray, reddish, greenish	S	Massive, crusts, stalactites, powdery	1–2.5	Neon yellow	Earthy	Opaque	White, clear, pale yellow	Production of fertilizers, sulfuric acid, insecticides, gunpowder, sulfur dioxide, etc.
Talc	Present but not often visible in hand sample	Pale green to dark green, greenish-gray, white, silvery-white, gray, brownish	$Mg_3Si_4O_{10}(OH)_2$	Crystals thin tabular or foliated or fibrous	1	Greasy and extremely soft	Pearly or dull	Translucent	White	Used as filler for paints, paper and rubber and in plasters, lubricants, toilet powder, chalk.

IGNEOUS ROCKS AND THE VOLCANOES THAT PRODUCE THEM

INTRODUCTION

Igneous rocks are the most abundant type of rock on Earth, when considered by volume. They all began as a liquid melt (magma) and cooled to form solid rock. Magma and therefore igneous rocks are brought to the surface by means of volcanoes. While volcanic activity brings life-giving nutrients and fresh crustal material to the surface for biological consumption, they also significantly endanger the surrounding area through the release of toxic gases, ash falls, lava flows, and lahars, among a number of other hazards. There is a common saying in geology—"the key to the present is the past." The igneous rocks and geological strata in the area surrounding volcanic activity are the history books for the volcano in question and in turn, when read properly, tell us what hazards the surrounding area is likely to face. Once the hazards and the distance to which they are a threat are identified, appropriate precautionary measures and mitigation techniques can be employed to minimize death and destruction.

Mt. St. Helens before Eruption

OBJECTIVE

The objective of this lab is twofold: 1) learn how to determine what hazards the area is likely facing based on the igneous rock assemblage and surrounding geology, and 2) use a case study of Mt. St. Helens and several other recently active volcanoes to demonstrate hazard assessment and mitigation.

Mt. St. Helens after Eruption

47

Igneous Rock Identification

Most, if not all, igneous rocks can be identified based on simple observational analysis by considering two properties of the rock: 1) texture, and 2) composition.

Texture – The external feel of the rock

The texture of an igneous rock varies depending on whether it cooled above the ground (**extrusive/volcanic**), and thus rapidly, or below the ground (**intrusive/plutonic**), and thus slowly. Rocks that cooled slowly below the ground are composed of large, coarse-grained minerals that can be seen and identified with the naked eye (**phaneritic**). Rocks that cooled rapidly above the ground are composed of small, fine-grained minerals that require the aid of a microscope to see and identify (**aphanitic**). Any composition of magma can therefore have at least two distinct textures and thus be classified as two different rocks.

Figure 4.1a

Figure 4.1b

── **Pegmatite Granite** ──

a. **Pegmatitic:** Very large crystals; grains range in size from several cm up to 30 ft in length—requires special circumstances to produce (e.g., granite pegmatite or pegmatitic granite). See Figures 4.1a and 4.1b.

Figure 4.2a – Granite

Figure 4.2b – Gabbro

b. **Phaneritic (Coarse-grained):** Visible grains, majority of crystals are uniform size and can be seen with the naked eye—the result of slow cooling below the surface (e.g., granite, diorite, gabbro). See Figures 4.2a and 4.2b.

Figure 4.3a – Basalt

Figure 4.3b – Rhyolite

c. **Aphanitic (Fine-grained):** Majority of crystals are very tiny and require a hand lens or microscope to be seen—the result of fast cooling at the surface (rhyolite, andesite, basalt). See Figures 4.3a and 4.3b.

Groundmass   Phenocrysts

Top: Figure 4.4a – Porphyritic Hand Sample
Bottom: Figure 4.4b – Microscope thin section image

d. **Porphyritic:** Two very contrasting sizes of crystals in the rock caused by slow cooling followed by a period of faster cooling. See Figures 4.4a and 4.4b.

- Phenocrysts: Larger crystals
- Groundmass or matrix: Smaller surrounding crystals

Figure 4.5 – Obsidian

e. **Glassy:** No mineral crystals develop due to instantaneous cooling (e.g., Obsidian—which is volcanic glass). See Figure 4.5.

f. **Vesicular:** Contains small holes or cavities called vesicles, which formed due to gas explosion in the lava as it rises and depressurizes. Typically porous and low density, it may resemble a sponge (e.g., pumice—light in color and low-density rock; scoria—usually brown, black or red in color and higher density rock). See Figures 4.6a and 4.6b.

Figure 4.6a – Pumice   **Figure 4.6b –** Scoria

Figure 4.7a – Volcanic Tuff

Figure 4.7b – Volcanic Breccia

Figure 4.7c – Volcanic Bomb

g. **Fragmental:** Rock contains broken, angular fragments of rocky materials welded together by heat during a volcanic eruption (e.g., volcanic breccia—welded together rock debris). See Figures 4.7a, 4.7b, and 4.7c.

What Story Does Texture Tell?

The texture of an igneous rock conveys where the magma/lava cooled, thereby specifying whether cooling took place underground, on the surface, or in the air. For example, fine grained rocks result from fast cooling at the surface, thereby indicating lava flow; whereas fragmental rocks weld together airborne and surface debris produced during an explosive eruption. As such, the texture is a key indicator as to whether the volcano produces surface hazards and if so, what type and to what extent.

Composition

The composition of an igneous rock is determined by identifying the minerals that make up the rock, which in turn is dictated by the chemical composition of the original melt from which the rock formed. The general composition of an igneous rock can be estimated by simply looking at the ratio of colors in the rock—felsic rocks generally contain lighter colored minerals (white to pink to light gray) and are less dense than mafic rocks, which tend to be darker in color (brown, black and dark green) as well as denser. See Figure 4.8.

- **Composition/Color Index**
    a. **Felsic/granitic:** Rock composed mainly of light-colored minerals like quartz and feldspar (dark minerals < 15 %) (e.g., rhyolite, granite)

    b. **Intermediate/andesitic:** Mixture of light- and dark-colored minerals like quartz, feldspar, amphibole, and biotite (dark minerals ~15%–45%) (e.g., andesite, diorite)

    c. **Mafic/basaltic:** Dark-colored minerals like olivine, pyroxene, and amphibole account for ~45%–85% of the rock. Plagioclase feldspar makes up most of the rest of the rock (e.g., basalt, gabbro)

    d. **Ultramafic:** Composed almost exclusively of dark-colored minerals like olivine and pyroxene (most commonly peridotite, which is found in the Earth's mantle)

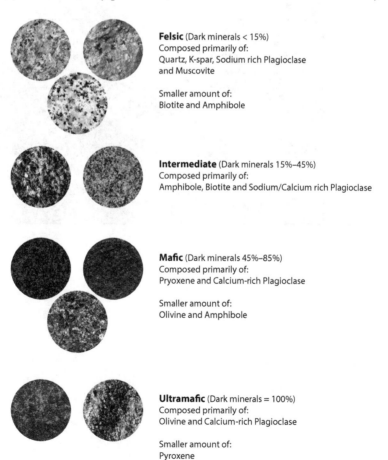

Felsic (Dark minerals < 15%)
Composed primarily of:
Quartz, K-spar, Sodium rich Plagioclase
and Muscovite

Smaller amount of:
Biotite and Amphibole

Intermediate (Dark minerals 15%–45%)
Composed primarily of:
Amphibole, Biotite and Sodium/Calcium rich Plagioclase

Mafic (Dark minerals 45%–85%)
Composed primarily of:
Pryoxene and Calcium-rich Plagioclase

Smaller amount of:
Olivine and Amphibole

Ultramafic (Dark minerals = 100%)
Composed primarily of:
Olivine and Calcium-rich Plagioclase

Smaller amount of:
Pyroxene

Figure 4.8 – Mineralogical Compositions of Igneous Rocks

What Story does Composition Tell?

The type of eruption that takes place and therefore the various hazards the surrounding area will face, is dictated primarily by the composition of the magma—specifically, the silica (Si) content. Si is the control on the magma's viscosity, i.e., resistance to flow. The more Si the magma has, the higher the magma viscosity and the more difficult it becomes for gas bubbles to escape the magma on its ascent to the surface. Thus, when magma contains a high concentration of Si, the gasses are forced to burst out of the magma, generating explosive volcanic activity. Exacerbating this effect is the fact that high Si magmas inherently contain significantly higher concentrations of dissolved gasses than lower Si ones and are generally lower temperature, which further reduces the fluidity of the magma. These additional factors substantially intensify the explosive nature of intermediate and felsic eruptions. On the flip side, mafic magmas, which contain lower Si concentrations, have little threat of explosion, but the high temperature, fluid lava can travel for miles resulting in significant damage to property. See Figure 4.9 for visual representation of this concept.

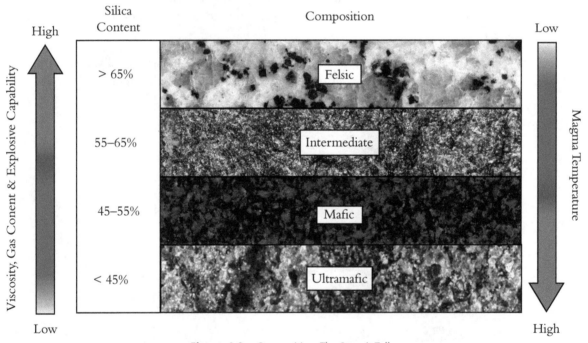

Figure 4.9 – Composition: The Story it Tells

How do we name Igneous Rocks?

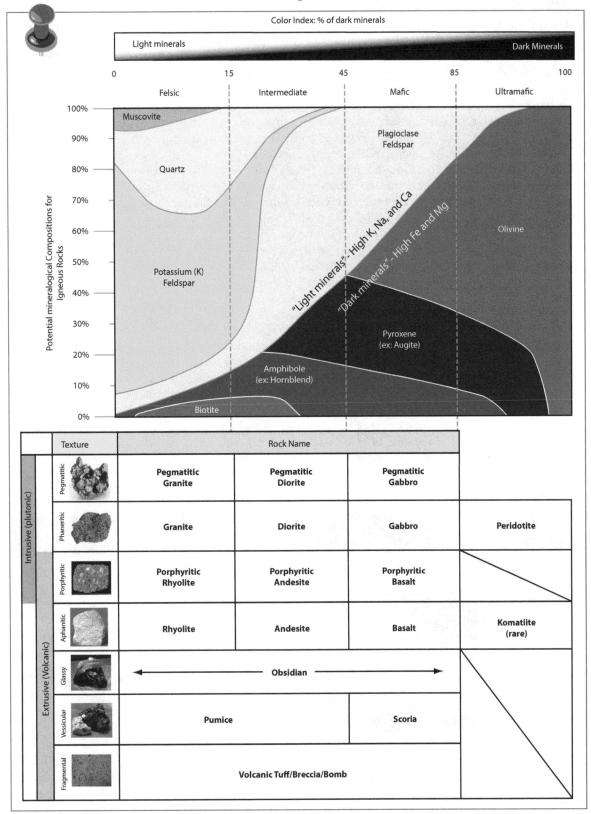

Figure 4.10 – Texture + Composition = Igneous Rock Name

Where do Volcanoes come in?

Volcanoes are vents in Earth's crust through which subsurface magma ascends and exits at the surface. Over the course of multiple eruptions, volcanic products (lava, ash, or both) accumulate around the vent, generating a volcanic cone that reflects the nature of the magma that produced it.

Mafic magmas, which have been brought up from the mantle and asthenosphere, have a relatively low Si concentration when compared to intermediate and felsic magmas. The lower Si content, in combination with the high temperature of the melt (the deeper the magma formation, the higher the temperature), generates low viscosity magmas that flow with ease and release gas bubbles with very little resistance. Because the primary extrusive product is low viscosity lava, the cone is built up from layer upon layer of thin lava flows, which create a wide, mildly sloping cone that mimics the shape of a warriors shield. Thus, volcanoes that erupt mafic magma are known as "Shield Volcanoes." Shield Volcano eruptions are considered nonexplosive and have little threat of hazards such as ash falls, pyroclastic flows, lateral blasts, or lahars. They do, however, cause significant problems with lava flows, as the fluid melt can travel long distances.

Felsic magmas—the result of shallow crustal melting—contain a high concentrations of Si. The elevated Si, in combination with the low magma temperature, produces a highly viscous magma, which traps gas bubbles and explodes violently upon depressurizing. In many cases, the lava is too viscous to flow any sizeable distance, making ash and other ejecta the primary extruded products. Thus, these volcanic cones are created from steep piles of consolidated ash that tightly hug the vent, creating the shape of a dome. As such, volcanoes that erupt felsic magma are known as "Volcanic Dome Volcanoes." Volcanic Dome eruptions are accompanied by ash falls, pyroclastic flows, lateral blasts, and lahars, yet lava flows are not an issue, as the lava is too viscous to travel any significant distance.

Intermediate magmas, the result of a combined contribution from mantle and crustal melting, produce magmas with a Si concentration and temperature between that of mafic and felsic magmas. As such, both lava and ash are extruded in comparable amounts, generating a steep cone built from alternating layers of viscous lava and steep ash piles. Thus, volcanoes that erupt intermediate magma are called "composite" or 'strato-volcanoes"—in reference to the alternating material that make up their cones. Because the lava erupted at composite volcanoes is of intermediate viscosity, it can and does flow a limited distance from the vent. Therefore, these volcanoes are associated with both lava flow problems, as well as violent, explosive eruptions, which produce all the hazards seen in felsic eruptions but to a lesser degree. See Table 4.A for a summary of this concept.

Table 4.A – Volcano Types and Their Behavior

Type	Shape	Si content	Magma Viscosity	Potential Eruption Products
Shield/Mafic	"Warriors Shield"	~50%	Low (fluid)	Lava flows, (ejecta that are kept extremely close to the vent)
Composite/Strato-Volcano/ Intermediate	"Conical"	~60%	Intermediate	Lava flows, gases, pyroclastic flows, ash falls, debris avalanches, lateral blasts, lahars (late stage highly siliceous lava sometimes produces a small lava dome)
Volcanic Dome/ Felsic	"Volcanic Dome"	~70%–75%	High (Sticky)	Lava dome, gases, pyroclastic flows, ash falls, debris avalanches, lateral blasts, lahars

See Table 4.B for a detailed description of select volcanic hazards associated with the three volcano types and their respective eruption styles.

Table 4.B – Volcanic Products and the Hazards They Produce

Products of Volcanic Eruptions	Visual Depiction	Creation process	Affected Area	Hazards Produced
LAVA ACTIVITY — Lava Flows	Pahoehoe / Aa	Non-explosive lava eruption from the vent.	Moves from high ground to low ground and can travel up to 10 km from the vent.	Unless on a steep slope, most people can easily outrun them. However, they can have a devastating effect on property, crops, and the landscape.
PYROCLASTIC ACTIVITY — Ash Fall		Bits of pulverized rock torn from the vent and glass created from instantaneous cooling of lava drops on contact with air. Particles < 2 millimeters (0.1 in) in diameter.	Covers hundreds and thousands of square km downwind from the source with volcanic ash.	Short-term destruction of vegetation and ecosystem, structural damage from increased load on roofs, respiratory and eye problems, "flame-out" in aircraft engines.
Pyroclastic Flow		Fast-moving current of hot gas and rock, created from destabilization of large masses of highly viscous lava. Can reach temperatures up to 2,000°F.	Travels at speeds up to 700 km/h (450 mph). downhill for several kilometers and may affect areas as far as 40km downslope.	Moving boulders will flatten trees and buildings and the hot gases will vaporize and set on fire much of the material in their paths.
Volcanic Bomb		Mid-flight solidification of lava into pieces larger than 65 mm (2.5 inches) in diameter.	Bombs can be thrown many kilometers from an erupting vent.	Volcanic bombs can cause severe injuries and death, as well as significant damage to property on impact.
Lateral Blast		Eruption that takes place on the flanks of a volcano instead of at the summit. Debris ejected at speeds of several hundred km/hr.	Distribution depends on direction of the blast and distance traveled depends on the strength of the explosion.	Explosive force can scorch and knock over upright structures (trees and buildings) several miles from the blast. Blocks/bombs can be thrown to distances of at least 10 km and pyroclastic flows, which can travel at high speed to distances of more than 30 km, can be produced.

Table 4.B – Volcanic Products and the Hazards They Produce

Products of Volcanic Eruptions	Visual Depiction	Creation process	Affected Area	Hazards Produced
LANDSLIDES — Debris Avalanche		Chaotic movement of rocks, soil, ash, vegetation, water, and/or ice on a volcanic slope. Can occur with *or* without an eruption.	Debris runs down the slope of the volcano and may extend as far as 45 km.	Rapidly moving mass of debris can flatten most obstacles in its path, as well as block roads and dam up rivers.
Lahar (mudflow)		Water, produced during eruption by lava melting an overlying glacier or in the absence of eruption by excess precipitation, mixes with pyroclastic debris generating a rapid flow that can reach depths of >100 feet and move at speeds of 100 km/hr.	Usually travel down valleys and can travel long distances—some have been known to travel hundreds of km from their source.	Can bury and destroy entire cities by covering cars, man-made structures (roads, bridges, and houses), and leaving tons of heavy debris in the trees.
ATMOSPHERIC HAZARD — Volcanic Gases		Explosive and non-explosive eruptions produce hot, toxic gases such as CO_2, CO, SO_2, H_2S, and HF, among other dangerous compounds.	Carried downwind at speeds of up to 10s of km/hr, depending on the strength of the wind.	Causes death by asphyxiation, damage via acidic corrosion, and water and soil contamination. Produces "Vog"—thick acidic haze made from a combination of volcanic gases, H_2O, and oxygen. Has been known to cause breathing problems, headaches, sore throats, watery eyes and flu-like symptoms.
HYDROLOGICAL HAZARD — Floods		Among other triggers, flooding can be caused by sudden introduction of excess water into a region or existing river from melting ice, by deposition of debris in a river that results in a substantial water-level increase, or a natural dam produced by debris blocking the normal river path.	Flooding may extend for 100s of km and is confined to flat areas.	Primary effects (those directly caused by the flood): injury, loss of life, damage caused by swift currents, debris, and sediment to farms, homes, buildings, roads, bridges, and communication systems, and erosion and deposition of sediment in urban and rural landscapes may also result in loss of soil and vegetation.

Tying it All Together – Volcanism and Plate Tectonics

While we've discussed how igneous rock composition can be used to understand the volcanic processes in a given area (see Figure 4.11), what we have not addressed is why magma composition—and thereby eruption style and products—differs from one locality to another. In a nutshell, the answer to that question is the Rock Cycle, which is driven by plate tectonics.

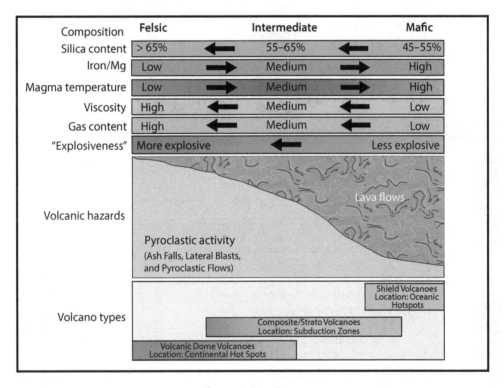

Figure 4.11 – Sum up

Throughout the rock cycle, there are three places where fresh magma rises to the surface: 1) In the middle of the ocean (spreading centers)—where two plates are pulling apart; 2) at subduction zones—deep oceanic trenches where the plates have collided; and 3) hot spots—localized volcanic areas that are not on plate boundaries. Refer to Figure 4.12 and corresponding Table 4.C to gain a better understanding of the way different magma compositions are produced at these various parts of the rock cycle. Then refer back to Figure 4.11 for a better overall understanding of why volcanoes in these different environments behave the way they do.

Figure 4.12 – Volcanic Settings and Magma Composition

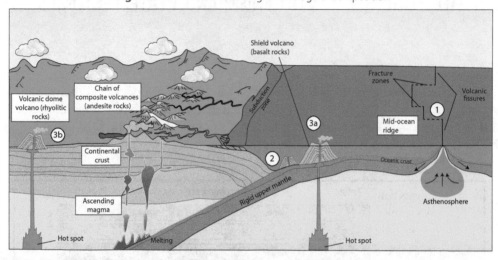

	Location of Volcanism	Cause of Magma Generation		Magma Composition	Volcano Type
Spreading Centers #1 in Fig. 4.12	Middle of Ocean Basins Where two plates pull apart	Opening of large fissures in the middle of the ocean releases the pressure on deeper, very hot, ultramafic rock, which then expands and rises		Ultramafic rock partially melts generating a *mafic* magma	Fissure eruptions
Subduction Zone #2 in Fig. 4.12	Over Oceanic Trenches Where two plates collide	Water is released from the subducting slab, moves up, and lowers the melting temperature of the upper mantle. Partial melting of the ultramafic uppermantle generates mafic magma. Insulation of mafic magma within the felsic continental crust generates magma mixing.		Range from borderline mafic-intermediate to borderline intermediate-felsic Constantly evolving	Composite
Oceanic Hot Spot #3a in Fig. 4.12	Mid Ocean above the Hot Spot	Hot, ultramafic rock rises from the core/mantle boundary due to thermal anomaly. Produces partial melting of ultramafic rock, which results in mafic magma.	Breaks through oceanic crust and forms volcanic island.	Mafic	Shield
Continental Hot Spot #3b in Fig. 4.12	On Continent above the Hot Spot		Stall beneath continental crust and partially melts intermediate rock, generating felsic material.	Felsic	Volcanic Dome

Table 4.C – Volcanic Settings and Magma Composition

Part A: Rock Identification and Hazard Assessment

Use the Igneous Rock Identification Chart (Figure 4.10), the hazard chart (Table 4.B), and the information about volcanic settings, (Figure 4.12 and Table 4.C) to answer the following questions. *If rock samples are unavailable, refer to the appendix at the end of the lab manual for photographic representations.*

Group A: Look carefully at the samples in this group. Notice that the rocks in this group all have coarse grains and a felsic composition.

- What is the name of the rocks in this sample group? _____

- Were they formed intrusively, extrusively, or both? _____

Group B: Look carefully at the samples in this group. Notice that the rocks in this group all have fine grains and a felsic composition.

- What is the name of the rocks in this sample group? _____

- Were they formed intrusively, extrusively, or both? _____

Questions

1. What minerals could potentially be found in these samples?

   _____

   _____

2. What type of volcano likely produced these rocks? **Shield, Composite,** or **Rhyolitic.**

3. Which of the following tectonic settings could have produced this volcano? (Circle all that apply) **Spreading Center, Subduction Zone, Oceanic Hotspot,** or **Continental Hotspot.**

4. Would the Si content of the magma in this volcano be **high, medium,** or **low**?

5. Would the viscosity of the magma from this volcano be **high, medium,** or **low**?

6. Would you expect this volcano to be dangerously explosive? **Yes** or **No**?

7. Now that you know the type of volcano that produced these rocks, determine whether the area in which these rocks were found is at high or low risk for the following hazards:

   a. Ash Fall: **High** or **Low**

   b. Lava Flow: **High**, **Medium**, or **Low**

   c. Mud Flows: **High** or **Low**

   d. Pyroclastic Flow: **High** or **Low**

   e. Lateral Blast: **High** or **Low**

Group C: Look carefully at the samples in this group. Notice that the rocks in this group all have coarse grains and an intermediate composition.

   • What is the name of the rocks in this sample group? _____

   • Were they formed intrusively, extrusively, or both? _____

Sample D: Look carefully at this sample. Notice this sample is porphyritic and has an intermediate composition.

   • What is the name of this rock? _____

   • Was it formed intrusively, extrusively, or both? _____

Questions

8. What minerals could potentially be found in these samples?

   _____

   _____

9. What type of volcano likely produced these rocks? **Shield**, **Composite**, or **Rhyolitic**.

10. Which of the following tectonic settings could have produced this volcano? (Circle all that apply) **Spreading Center**, **Subduction Zone**, **Oceanic Hotspot**, or **Continental Hotspot**.

11. Would the viscosity of the magma from this volcano be **high**, **medium**, or **low**?

12. Would the Si content of the magma in this volcano be **high**, **medium**, or **low**?

13. Would you expect this volcano to be dangerously explosive? **Yes** or **No**?

14. Now that you know the type of volcano that produced these rocks, determine whether the area in which these rocks were found is at high, medium, or low risk for the following hazards. Assume the area has a reasonably high precipitation rate.

    a. Ash Fall: **High** or **Low**

    b. Lava Flow: **High**, **Medium**, or **Low**

    c. Mud Flows: **High** or **Low**

    d. Pyroclastic Flow: **High** or **Low**

    e. Lateral Blast: **High** or **Low**

Sample E: Look carefully at this sample. Notice this sample is coarse grained and has a mafic composition.

- What is the name of this rock? _____

- Was it formed intrusively, extrusively, or both? _____

Sample F: Look carefully at this sample. Notice this sample is fine grained and has a mafic composition.

- What is the name of this rock? _____

- Was it formed intrusively, extrusively, or both? _____

Questions

15. What minerals can you identify in these samples?

_____

_____

16. What type of volcano likely produced these rocks? **Shield**, **Composite**, or **Rhyolitic**.

17. Which of the following tectonic settings could have produced this volcano? (Circle all that apply) **Spreading Center**, **Subduction Zone**, **Oceanic Hotspot**, or **Continental Hotspot**.

18. Would the Si content of the magma in this volcano be **high, medium,** or **low**?

19. Would the viscosity of the magma from this volcano be **high, medium,** or **low**?

20. Would you expect this volcano to be dangerously explosive? **Yes** or **No**?

21. Now that you know the type of volcano that produced these rocks, determine whether the area in which these rocks were found is at high, medium, or low risk for the following hazards. Assume the area has a reasonably high precipitation rate.

    a. Ash Fall: **High** or **Low**

    b. Lava Flow: **High, Medium,** or **Low**

    c. Mud Flows: **High** or **Low**

    d. Pyroclastic Flow: **High** or **Low**

    e. Lateral Blast: **High** or **Low**

Just for Show

Go to the counter in the lab and take a quick look at some other products volcanoes are capable of producing.

Sample W: This is an example of a pegmatite. Pegmatites are extremely unusual rocks for a number of reasons, not the least of which because they form deep in the lithosphere and are comprised of abnormally large grains, but are actually the result of unusually fast cooling. These oddities are created from the very end stages of felsic magma crystallization, where all the water in the magma has pooled, permitting rapid movement of the atoms and therefore rapid growth of crystals. Pegmatite crystals usually range in size from several centimeters to up to 10 m in length. This is the one case where significant crystal growth occurs in a very short period of time. Pegmatites are also the home of a variety of "odd" elements, known as "rare earth metals," which include all the lanthanides on the periodic table, as well as Scandium and Yttrium. Rare earth metals do not comfortably fit into most minerals and are forced to collect in the remaining magma. As such, pegmatites contain high concentrations of rare earth metals, which are essential to industry, particularly the electronics field. Notice the size of the crystals relative to the standard course grained rock.

- How might the prevalence or scarcity of these rocks affect political relations between countries? _____

_____

_____

_____

Sample X: This is a sample of what is known as "obsidian" or "volcanic glass." This material forms when a felsic magma is so viscous, the atom are unable to move and therefore bond with the closest atom instead of creating an organized crystal structure. The resulting material is therefore full of randomly bonded atoms, which creates a glass. Obsidian was commonly used by the Native Americans to make arrowheads and today is frequently used in jewelry.

- Is this rock composed of minerals? **Yes** or **No**?

  Explain: _____

  _____

  _____

Sample Y: This rock forms when a fragment of magma is spat into the air and cools while in midflight. Because it is fighting air resistance while cooling, these rocks have a football or teardrop shape. For those of you who saw the movie Dante's Peak, this rock is what killed Pierce Brosnan's girlfriend at the very beginning of the movie!

- What is this rock called? _____

Sample Z: These are samples of Pumice (light-colored one) and Scoria (dark-colored one). These rocks form from exploding gas bubbles attempting to escape the magma as it approaches the low pressure surface. The escaping gasses stir the lava into a frothy foam, which cools very quickly upon contact with atmosphere. The holes that you see are where gas bubbles were exploding when the froth cooled. The process is very similar to what happens when you take the top off a soda bottle that has been shaken up.

- The felsic pumice cools instantaneously, thereby forming a glass, whereas the mafic scoria cools slower, forming a fine grained, extrusive rock. Explain why this is the case.

  _____

  _____

- Why does pumice make a good skin exfoliate? _____

  _____

Part B: Mt. St. Helens Case Study

The Cascadia subduction zone marks a convergent boundary where the Juan de Fuca plate is subducting beneath the over-riding North America plate (See Figure 4.13a). The fault runs from northern Vancouver Island to northern California and is bordered on the northern and southern ends by triple junctions (See Figure 4.13b). In addition to being the source of some significant earthquakes, the subduction zone is also home to the Cascade Volcanic Arc, a chain of stratovolcanoes averaging 10,000 feet high that extend from northern California to British Columbia. The major peaks from south to north include: California's Lassen Peak and Mt. Shasta; Oregon's Crater Lake (Mazama), Three Sisters, Mt. Jefferson, and Mt. Hood; Washington's Mt. Adams, Mt. St. Helens, Mt. Rainier, Glacier Peak, and Mt. Baker; and British Columbia's Mt. Garibaldi and Mt. Meager.

Mt. St. Helens is located in southwestern Washington State and is a typical Cascade volcano. Events that occurred and hazards generated by this volcano during the 1980 eruption would likely be repeated if other Cascade volcanoes were to erupt. Mt. Rainier is close to the Seattle-Tacoma area of Washington, and Mt. Hood is close to the Portland area of Oregon (see Figure 4.14). Due to the high population density on the west coast, eruptions of any Cascade volcano might have major impacts on the nearby metropolitan areas.

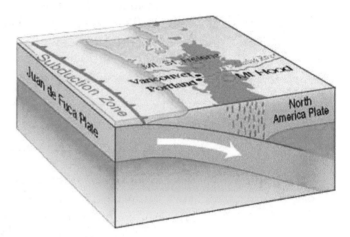

Figure 4.13a

Figure 4.13b

Figure 4.13 – Cascadia Subduction Zone

There were several distinct events at Mt. St. Helens on the morning of May 18, 1980. Among them were 1) an earthquake of Richter magnitude 5.1, 2) a landslide on the north side of the mountain, 3) a lateral blast to the north, 4) a vertical eruption column, 5) lahars down many of the river valleys, and 6) extensive ash fall carried all the way to the east coast of the US by westerly (blowing from the west) winds. Let's now examine the extent to which these hazards affected the surrounding area and use that knowledge to determine necessary safety measures to prevent injury and loss of life and property in the event another Cascade volcano erupts.

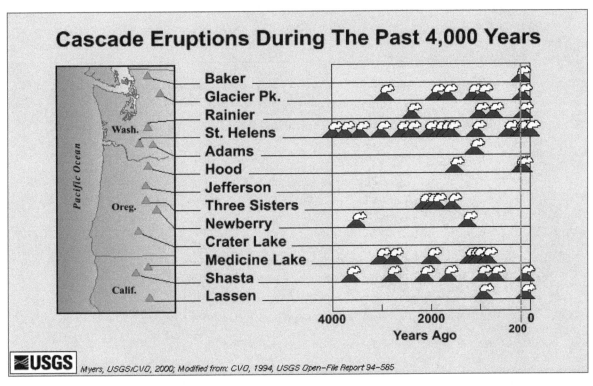

Figure 4.14 – Major Cascade Volcanoes on the West Coast of the U.S. and their Eruption Frequency.

Mt. St. Helens: Topography before and after: Use *Maps 4-I and 4-II* to answer the questions below.

1. Map 4-I is a topographic map of Mt. St. Helens prior to the eruption and Map 4-II is a map of the mountain after the eruption. A traverse line bisecting the mountain has been drawn on both maps. Enlarged views of the traverse region are found on page 69.

    *   Using the close-up view, i.e., Figure 4.15a, draw a topographic profile of the A →B traverse line on the graph paper provided. *You only need to use index contour lines to do your profile.*

    *   Using the second piece of graph paper provided, repeat the above procedure, this time drawing a topographic profile of the C →D traverse line on the close up view of Mt. St. Helens (Figure 4.15b) provided from Map 4-II.

2. On the bullet points below, list the 2 major differences between these 2 profiles.

    *   _____

    *   _____

3. At the point of maximum elevation loss, how many feet of elevation were lost from the mountain? _____ ft

4. From the shape of the mountain prior to eruption, is Mt. St. Helens closer in profile to Hawaiian shield volcanoes such as Mauna Loa and Mauna Kea or to stratovolcanoes such as Mt. Shasta?

    _____

5. Based on the shape, deposits, and location of the volcano, what composition of rock would you expect to find here? **Mafic**, **Intermediate**, or **Felsic**.

6. Notice there are areas on or near the volcano where elevation increased instead of decreased.

    *   What do you think caused this increase in elevation?

    _____

    _____

    *   What hazard becomes a greater concern as a result of this increase in elevation?

    _____

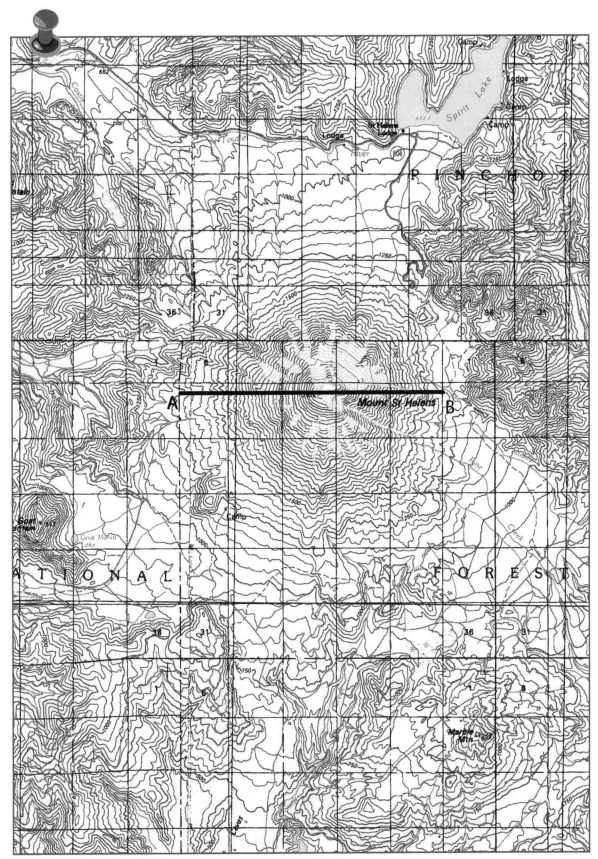

Map 4-1 – Topography of Mt. St. Helens Prior to Eruption

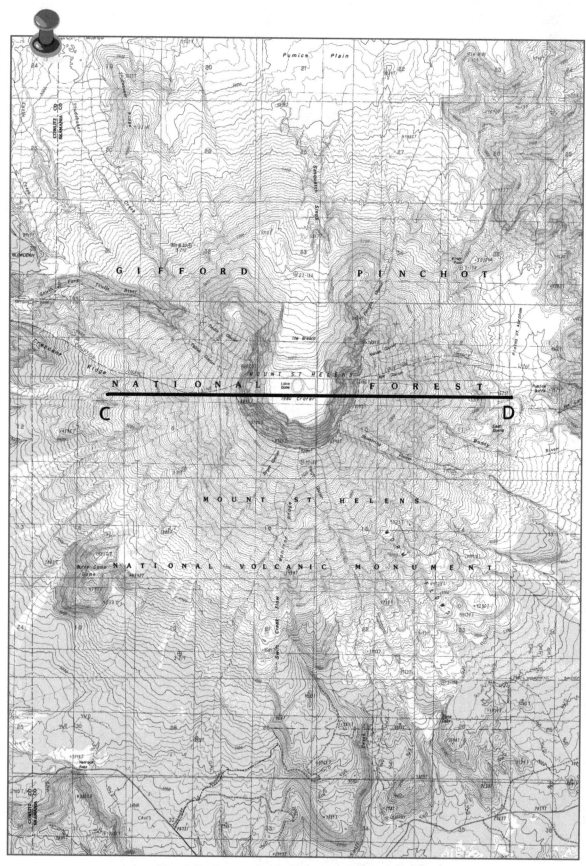

Map 4-II – Topography of Mt. St. Helens after the Eruption

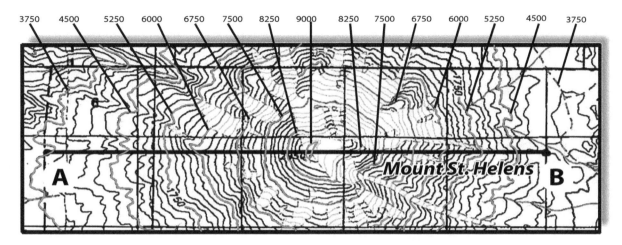

Figure 4.15a: A–B Traverse Line on Map 4-I

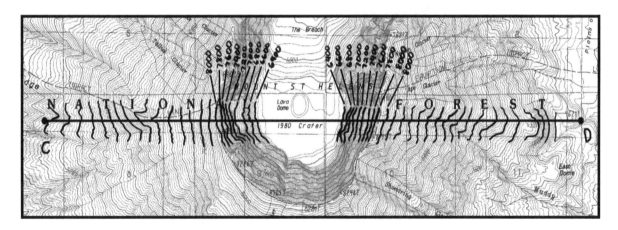

Figure 4.15b: C–D Travese Line on Map 4-II

★Note: Because Map 4-I was constructed in meters, the elevations in bold on this enlargement have been converted to feet so they can be appropriately compared to the profile from Map 4-II.

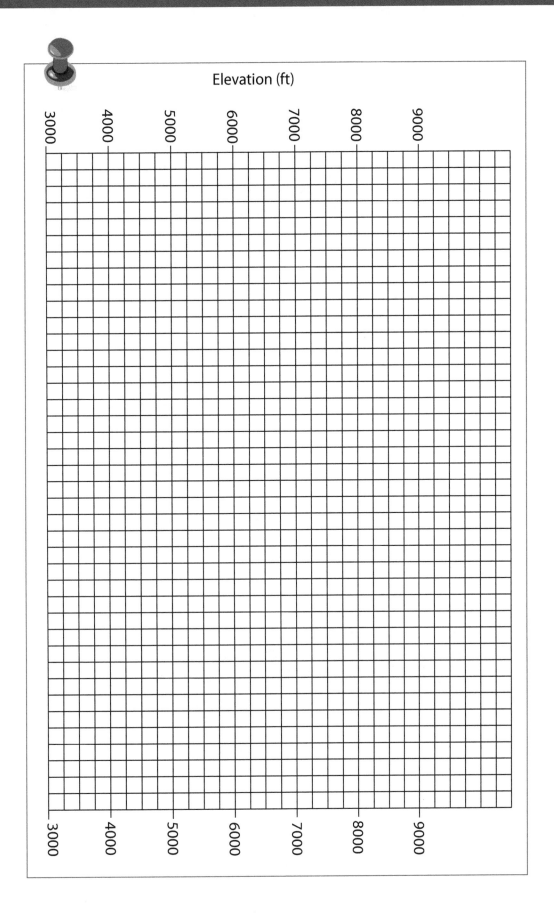

Elevation (ft)

3000 4000 5000 6000 7000 8000 9000

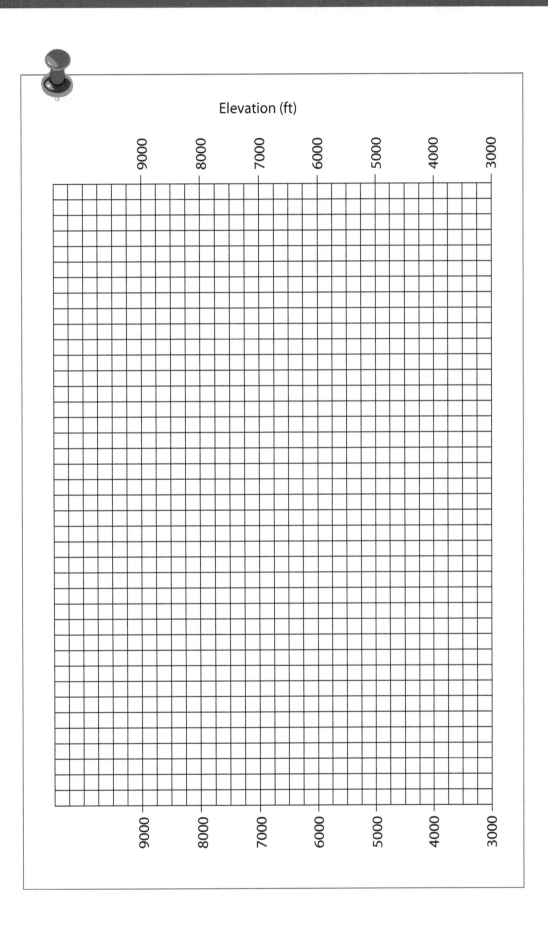

Elevation (ft)

7. Examine Maps 4-I and 4-II to determine the distribution of glaciers near the top of Mt. St. Helens both before and after the eruption.

    • What changes took place? _____

    • Where did the ice go? _____

    • What two serious hazards were created as a result? _____ & _____

Volcanic Deposits and their Locations: Figure 4.16 below as well as Figures 4.17–4.19 depict various volcanic deposits from the 1980 eruption of Mt. St. Helens.

8. Which of the deposits are found exclusively within a 0–20-mile radius of the volcano—"close," and which were able to get significantly farther than >20 miles from the volcano—"far away"? On the lines below, label each hazard as either "close" or "far away."

    • Pyroclastic flow deposits _____

    • Mud flow deposits _____

    • Debris—Avalanche deposits _____

    • "Scorch Zone" _____

    • Ash fall deposits _____

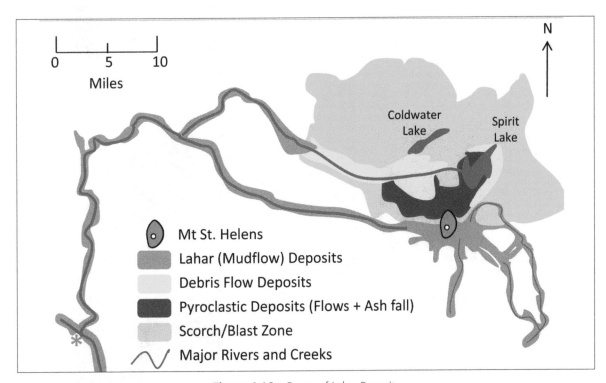

Figure 4.16 – Extent of Lahar Deposits.

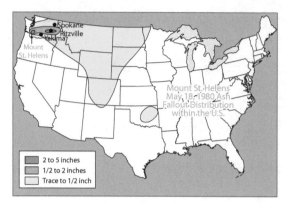

Figure 4.17 – National Ash Fallout Deposits.

Side note: This image depicts the ash cloud distribution from the April 17, 2010 volcanic eruption beneath the Eyjafjallajokull glacier in Iceland, which grounded planes in Europe for nearly a week.

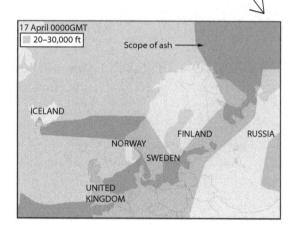

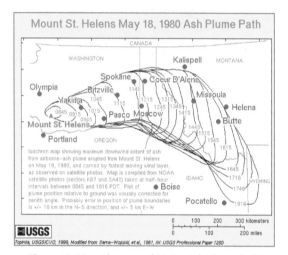

Figure 4.18 – Close-up of Major Ash Plume Path.

Figure 4.19a – View of downed Tree Zone at Mt. St. Helens

Figure 4.19b – Pyroclastic Flow on Mt. St. Helens

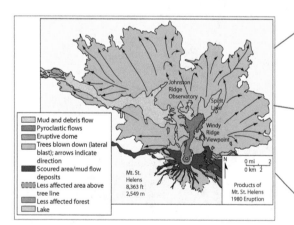

Figure 4.19 – Volcanic Deposits from 1980 Eruption.

Figure 4.19c – Lahars from Mt. St. Helens

Figure 4.20 below shows the configuration of the Columbia River bed before and after the 1980 eruption of Mt. St. Helens.

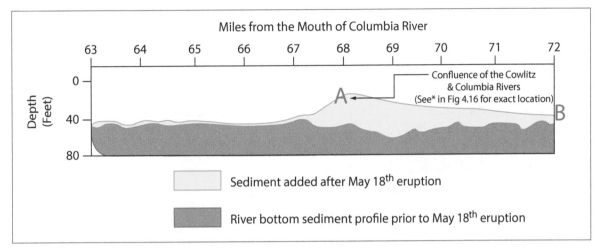

Figure 4.20 – Bed-Load profile of the Columbia River before and after the May 18th, 1980 eruption.

9. Beginning at the confluence of the rivers (Point A) and ending at Point B, calculate the average number of feet the river bed was raised? _____ ft (*Hint:* Measure the increase in height on the far right side, right at the confluence and the area in between, then average those measurements.)

10. Is this going to increase or decrease the magnitude of flooding? _____
    Explain: _____
    _____

ASSESSING THE LIKELY HAZARD

11. If we hypothesize that the other Cascade volcanoes will erupt in a similar fashion, suggest ways for surrounding communities to protect against the following:

    a. Ash Falls _____

    b. Floods _____

    c. Lahars _____

The only way to keep safe from pyroclastic flows, lateral blasts, lava flows, and volcanic gases is simply to stay out of their path. Examine Figures 4.16 and 4.19 one more time.

12. What radius from the volcano, in miles, would you designate as the "High Risk Zone"—i.e., "Do not build here"? _____ miles

13. In any problematic location, i.e., where an active volcano exists, given the propensity for pyroclastic and debris flows to move downhill, lahars to infiltrate the rivers, and ash and volcanic gases to be distributed by the wind, where would you suggest the city be built with respect to the volcano concerning:

    a. Topography: Upslope or downslope from the volcano? _____

    b. Rivers: Close to or significantly away from river systems? _____

    c. Wind direction: Upwind or downwind from the volcano? _____

14. How likely is it that all the criteria you determined in #13 as necessary to safely build in a volcanic setting will be met at any one location? Explain.

    _____

    _____

    _____

SEDIMENTS, SEDIMENTARY ROCKS, AND SLOPE STABILITY

INTRODUCTION

Sedimentary rocks make up approximately 5% of the Earth's crust and < 1% of bulk earth by volume, yet they compose more than 75% of Earth's surface. Despite the fact that they are a minority, they play a critical part in the sustenance of the biosphere. The weathering processes that create sediments transforms the nutrients brought up by igneous processes into a form that can be taken up by organisms and thereby incorporated into the biological system. It is also sedimentary processes that continuously move materials and nutrients from the high ground to the low ground, bringing sustenance and resources to all parts of the planet. The constant movement of sediments, and therefore nutrients and resources, both on the continents and in the ocean, can be accomplished through "Mass Wasting." Mass wasting, aka landslides, is the downslope movement of sediment and bedrock due to the influence of gravity. Because most of the surface is composed of sediment and sedimentary rock, most landslide hazards involve these materials. A basic understanding of how sediment and sedimentary rocks behave on slopes will allow identification of potentially dangerous areas (see cartoon below) and the development of appropriate precautionary measures and mitigation techniques to minimize death and destruction from this natural and necessary process.

OBJECTIVE

The objective of this lab is threefold: 1) examine some of the factors that impact the type and frequency of mass wasting; 2) learn how to determine what specific mass wasting hazard(s) an area is likely facing based on the sediment, sedimentary rock assemblage, and geological and meteorological conditions; and 3) evaluate potential preventative measures and mitigation techniques.

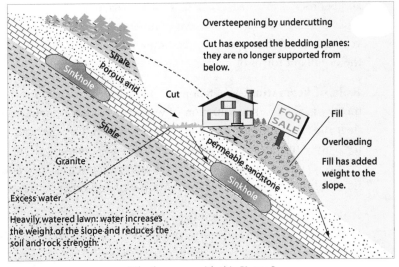

What's wrong with this Picture?

Slope Stability and Instability

Slopes are the most common landforms on Earth. Although slopes appear stable and static, they are in fact dynamic, evolving systems that will go through periods of stability separated by periods of movement (although there are some that are in constant, though very slow motion). The stability of a slope expresses the relationship between "driving forces" (moves materials down a slope) and "resisting forces" (opposes such movement). The most common driving forces are the weight of the slope material and the pull of gravity, while the most common resisting forces are friction and cohesion. Slope stability is determined by the ratio of the resisting forces to the driving forces. As local conditions change, the slope conditions change, increasing or decreasing the likelihood of slope failure. Driving and resisting forces are determined by the interrelationships of the following variables: type of earth materials, slope angle and topography, climate, vegetation, water, and time.

A. **Type of materials:** Affects both the type and the frequency of downslope movement. Unconsolidated deposits★ will move downslope more easily than bedrock.

   ★Note: Unconsolidated deposits include both sediment and soil. Sediment consists primarily of weathered mineral particles, whereas soil is sediment with the addition of substantial organic matter—the result of intense physical and chemical interaction with Earth's biosphere. An additional difference between sediment and soil is that sediment is transported away from its source of origin, while soil remains and accumulates in the vicinity of its parent rock.

   While unconsolidated deposits mass waste in the form of slumps and flows, most bedrock mass wastes by means of sudden slides and falls. In addition, slopes formed of weak, poorly cemented rock—such as shale, undergo very different types of mass wasting from slopes composed of strong, well-cemented rock—like sandstone or limestone. In addition, a third category of problems arise when weak, poorly-cemented rock is in between or alternates with layers of strong, well-cemented sedimentary rocks.

B. **Slope angle/topography:** The steeper the slope, the stronger the driving forces and the more frequent and intense the mass wasting.

C. **Role of climate:** Influences the amount and timing of precipitation and the type and abundance of vegetation that grows there. In arid and semi-arid regions, vegetation is sparse, soils are thin, and bedrock is exposed. Rock falls, rock avalanches, slides, and shallow soil slips are the common types of mass wasting in these environments. In humid and semi-humid regions, there is abundant vegetation and thick soil cover, resulting in complex landslides, flows, slumps, and soil creep being the dominant forms of mass wasting.

D. **Role of vegetation:** Vegetation provides cover that cushions the impact of falling rain, and thereby facilitates the infiltration of water into the slope, while slowing down erosion. In addition, their root systems open up routes through which the water can efficiently travel, thereby preventing the water from pooling, and also add cohesion to unconsolidated material. However, because the plants uptake much of the water themselves, they also add significant weight to the slope, which can result in facilitating soil slips and other forms of mass wasting.

E. **Role of water:** When it comes to slope stability, water is a double-edged sword. Under the right conditions and in the right concentration, water will add cohesion to loose particles and thereby help stabilize the slope by decreasing the frequency and magnitude of mass wasting. It is this same cohesive force that holds sand castles together (see Figure 5.2). However, much more often than not, there is too little water, which results in the absence of a cohesive force (see Figure 5.1), or too much water, such that the particles are semi-suspended in the fluid (see figures 5.3 and 5.4). These extremes facilitate mass wasting. In addition, if the rock body is not "permeable"—capable of efficiently transporting water through the rock—water can also lubricate the boundary between rock layers increasing the frequency of slides, and quietly build up inside slopes composed of rocks that store as opposed to transport the water, resulting in severe mass wasting even many months after a heavy rain or rainy season.

Figure 5.1 © ISTOCKPHOTO LP

Figure 5.2

Figure 5.3 © ISTOCKPHOTO LP

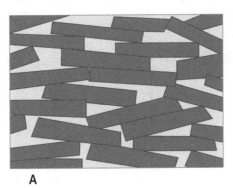

A

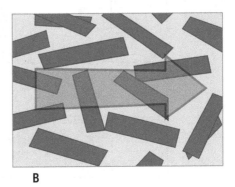

B

Figure 5.4

F. **Time:** Because the presence or absence of mass wasting is dependent on so many external factors, what's safe and stable today may not be so tomorrow. That's why if an area has even the potential to cause destructive landslides, adequate mitigation techniques should be implemented prior to development.

How do we classify mass wasting processes (see Figure 5.5)?

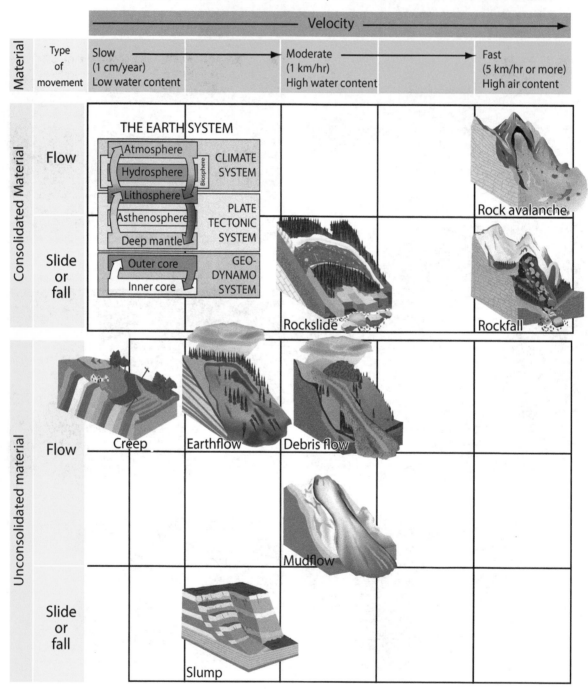

MASS MOVEMENTS ARE CLASSIFIED ACCORDING TO THE DOMINANT
MATERIAL, WATER OR AIR CONTENT, AND VELOCITY OF THE MOVEMENT

Figure 5.5 – Types of Mass Wasting

Table 5.A on the following page lists and defines the main type of mass wasting and the subdivision within each category.

Table 5.A –Types of Downslope Movement

Category	Movement	Characteristics
Slide	Translational Slide (Rock Slide)	Large block of bedrock or cohesive material slides on a planar surface
	Rotational Slide (Slump)	Cohesive material slides on a curved surface
Flow	Creep	Geologic surface material that flows downslope at an imperceptible rate
	Earth flow	Unconsolidated fine grained material that flows downslope in a partially fluid mixture
	Debris flow	Unconsolidated coarse and fine grained material that flows downslope in a saturated, fluid mixture
	Mudflow	Fine grained material that flows downslope in a saturated, fluid mixture
	Rock Avalanche	Combination of rocks, soil, vegetation, snow, and ice that moves rapidly downslope
Fall	Rock Fall	Rocks or blocks of rocks in freefall

Part A: Sediments and their Properties

Weathering causes pre-existing rock to break down into particle-sized pieces (clasts) or atoms dissolved in water (ions). These clasts and ions are known as **sediment** and are transported by water flow, wind, glaciers, and mass wasting to other localities. Following deposition, they will eventually undergo either lithification (if the sediment is clasts) or precipitation (if the sediment is ions), thereby forming sedimentary rock. The transformation of pre-existing rock to sedimentary rock takes place over thousands of years and therefore at any given time, much of Earth's surface is actually covered with unconsolidated material. Whether the loose clasts will remain stable or undergo mass wasting depends on the "angle of repose" for the slope and materials in question. The angle of repose is the steepest slope that a pile of unconsolidated sediment can have and remain stable. Factors that affect a slope's angle of repose are as follows:

A. **Grain Size:** Larger grains have more surface area with which to create friction and cohesion. As such, larger grains are more resistant to the driving forces and therefore slopes with course material have a higher angle of repose than those composed of fine-grained material (see Figure 5.6).

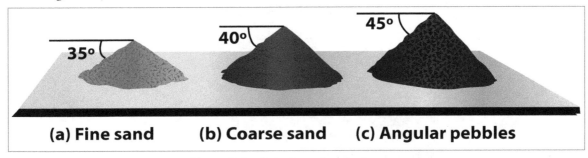

Figure 5.6 – Grain size and angle of repose.

B. **Grain Shape:** Sediment grains can be angular, sub-angular, sub-rounded, and well-rounded. These terms refer to the extent to which the corners of individual grains have been rounded off. Because angular grains have sharp corners, they are able to poke their corners through spaces between other grains, thereby creating a loosely locked network with more points of friction and cohesion, which is considerably more resistant to the driving forces than slopes composed of more rounded grains. Slopes with a high percentage of angular grains have a higher angle of repose than those composed of primarily rounded grains (see Figure 5.7).

C. **Grain Sorting:** The sorting of particles in a sediment refers to the range of particle sizes that are present. A well-sorted sediment contains particles of only one size. A poorly sorted sediment contains particles of varying sizes. Transportation by wind and rivers tends to produce well-sorted sediment, while transport by ice, such as in glaciers, commonly produces the most poorly sorted sediments with particle sizes ranging from mud to boulders, all deposited simultaneously (see Figure 5.7). Poorly sorted sediment has a higher angle of repose than well-sorted sediment, as the small grains fill in the holes between the larger ones, therefore increasing points of contact for friction and cohesion.

Unconsolidated Sediment and Landslides. Use what you know about the properties of unconsolidated sediment to answer the following questions.

1. On your table you will find sediment samples A, B, and C. Sediment Sample A is gravel, sediment Sample B is coarse sand, and sediment Sample C is fine sand. All three samples are well sorted and contain sub-angular grains.

    a. Of these three samples, which would you expect to be the most stable? _____

    b. Of these three samples, which would you expect to be the least stable? _____

    c. What property did you use to determine the answers to questions **a** and **b** above?

    _____

    d. How would the property you listed in question **c** affect the stability of the slope?
        affects _____ and _____

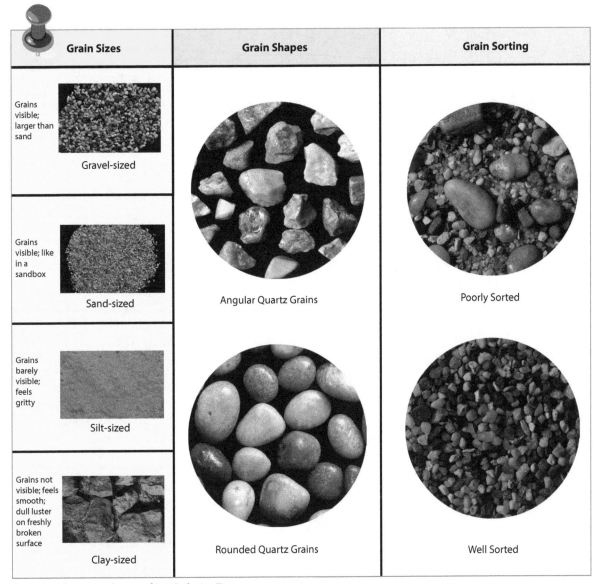

Grain Sizes	Grain Shapes	Grain Sorting
Grains visible; larger than sand — **Gravel-sized**	**Angular Quartz Grains**	**Poorly Sorted**
Grains visible; like in a sandbox — **Sand-sized**		
Grains barely visible; feels gritty — **Silt-sized**	**Rounded Quartz Grains**	**Well Sorted**
Grains not visible; feels smooth; dull luster on freshly broken surface — **Clay-sized**		

* Photographs are compliments of Dr. Catherine Forster.

Figure 5.7 – Intensity and frequency of mass wasting will in part depend on grain size, shape, and sorting.

2. Look at the column titled "Grain Sorting" in Figure 5.7. Examine the two photographs of the poorly–sorted and well–sorted sediment samples.

    a. Of these samples, which would you expect to be the most stable? _____

    _____

    b. Of these samples, which would you expect to be the least stable? _____

    _____

c. Explain your reasoning for answering questions **a** and **b** as you did:

_____

_____

_____

3. Look at the column titled "Grain Shapes" in Figure 5.7. Examine the two photographs of the angular and well-rounded sediment samples.

a. Of these samples, which would you expect to be the most stable? _____

b. Of these samples, which would you expect to be the least stable? _____

c. Explain your reasoning for answering questions **a** and **b** as you did:

_____

_____

_____

4. On your table you will find sediment samples D, E, and F. Sediment Sample D is dry sand, sediment Sample E is damp sand, and sediment Sample F is saturated sand. Examine the three samples closely, then answer the following questions.

a. Of these three samples, which has the most strength? _____

b. Why are the other two samples so much weaker? (Explain your reasoning for each, one at a time—the answer is not the same for both.)

Sample _____: _____

Sample _____: _____

c. Based on your answers to the previous questions, is it environmentally sensible to blast through miles of mountains, thereby cutting everything at the same angle, in order to build lengthy highways? Yes or No? Explain.

_____

_____

_____

If no, list at least 5 variables that should be examined at each location in order to determine the steepest angle at which the slope could be cut, but still remain stable.

- _____

- _____

- _____

- _____

- _____

5. Examine Figure 5.8. You are a coastal engineer asked to infill the eroding cliff beneath the road with boulder-size rocks in hopes of preventing the waves from removing any more material and further endangering the road. Given the answers you chose to the questions above, what grain shape (**angular** or **well-rounded**) and sorting (**poor** or **well**) would you use? (Circle the appropriate answers.)

Figure 5.8

Part B: Sedimentary Rocks and their Properties

Sedimentary Rocks are the most abundant type of rock on the surface of the Earth. They are composed of sediment—products of weathering consisting of loose rock fragments (detritus), clasts broken down from once-intact rock, ions that precipitate directly out of water, and biological debris. Sedimentary rocks have either accumulated or precipitated at the surface of the Earth and hold the key to understanding much of Earth's surface history, including the history of life on this planet. Sedimentary rocks can be divided into four categories: Clastic/Detrital, Chemical, Biochemical, and Organic. The primary differences between these types of sedimentary rocks concern the way the original rock was weathered (physical vs chemical), the type of sediment created (clasts vs ions), and the processes that ultimately form the sedimentary rock (lithification vs. precipitation).

Clastic/Detrital Sedimentary Rocks

All clastic rocks are composed of cemented particles (clasts) that have been subjected to moderate to intense physical and chemical weathering. Physical weathering breaks and chips larger pieces into clasts, while chemical weathering strips ions away from the outer layers of the rocks' minerals. The clasts that are transported (either by wind, water, or ice) away from the location where they were originally created are then deposited and lithified by means of compaction and subsequent cementation. Clastic sedimentary rocks are all composed of rock fragments of varying size, color, and composition and are differentiated based on the dominant grain size composing the rock (see Table 5.B).

Table 5.B – Clastic/Detrital Sedimentary Rocks

Sediment Type	Particle Size (mm)	Grain Shape	Rock Name	
Gravel or pebble	> 2	Rounded	Conglomerate	
		Angular	Breccia	
Sand	2–1/16	Angular to rounded	Sandstone	
Silt	1/16–1/256	N/A	Siltstone	
Mud	< 1/256	N/A	Shale (laminated and/or fissile—breaks into thin layers)	Mudstone
		N/A	Mudstone (non-laminated and non-fissile)	

Chemical Sedimentary Rocks

Chemical rocks are all created through chemical reactions and have been subjected to little if any physical weathering or erosion. Because their crystals were all created via chemical reactions, the primary diagnostic property is the chemistry of the rock (as opposed to the size of the particles, as was seen in clastic rocks). See Table 5.C for naming of select chemical sedimentary rocks.

Table 5.C – Chemical Sedimentary Rocks

Composition	Mineralogy	Rock name
Calcite (Fizzes with HCl)	Calcite	Crystalline limestone
Non-calcitic	Halite	Rock salt
	Gypsum	Rock gypsum
	Quartz	Chert/Flint
	Kaolinite and/or other clay minerals	Claystone

Biochemical and Organic Sedimentary Rocks

Some chemical rocks are produced with the aid of living organisms (plants or animals) and we refer to these as **biochemical** or **organic chemical rocks.** Rocks that are formed in this way require a combination of chemical and physical weathering processes to create the rocks. See Table 5.D for naming of select biochemical rocks and Table 5.E for naming of select organic rocks.

Note: Biochemical rocks are composed of the hard parts of once living organisms (shells and skeletons), while organic rocks are made form the soft tissue of once living organisms. This is why we classify them differently.

Table 5.D – Biochemical Sedimentary Rocks (limestones)

Composition	Particle Constituents	Rock name
Calcite (Fizzes with HCl)	Cemented mass of loose, coarse shell fragments	Coquina
	Mix of course and fine-grained fossil debris	Fossiliferous limestone (Fossils + micrite)
	Mostly very fine-grained to microcrystalline detrital mass of calcite and microfossils	Micrite
	Mostly fine-grained, earthy, chalky, light-colored mass of coccolithophore skeletons	Chalk

Table 5.E – Organic Sedimentary Rocks

Composition	Particle Constituents	Rock name
Carbon: The remains of vegetation that was buried before decay	Carbon particles	Coal

Sedimentary Rocks and Landslides

6. Find Sample X and set it next to the picture of poorly sorted sediment in Figure 5.7
    Next, find Sample Y and set it next to sediment Sample B (coarse sand)
    Lastly, find Sample Z and set it next to sediment Sample G (mud)

Sample X is a conglomerate/breccia, the rock formed when poorly sorted sediment gets lithified; Sample Y is a sandstone, the rock formed when sand gets lithified; and Sample Z is a shale, the rock formed when mud gets lithified.

    a. What two processes took place that transformed the weak and easily destabilized unconsolidated sediments into the relatively strong samples X, Y, and Z?
    _____ and _____

    b. Of samples X, Y, and Z, does any one appear weaker than the others? **Yes** or **No**?

    c. If yes, which one?_____
    Do you think this type of rock might be more prone to mass wasting problems than the others? Yes or No?
    Explain: _____
    _____
    _____

7. Examine Figure 5.9. This picture was taken in Luxor, Egypt; the cliff behind Hatshepsut Temple is composed of limestone. Limestone can have either a chemical origin (as in the case of crystalline limestone) or biochemical origin (as in the case of coquinas, fossiliferous limestones, chalk, and micrite); but in either case, limestone forms primarily from dissolution and precipitation processes. Limestone is a very common rock in the US, particularly in Florida and the Midwest. As such, many towns and cities are built on this material and therefore have to contend with the mass wasting problems it produces. After examining the photograph, answer the questions that follow.

    d. Look carefully at Figure 5.9. Does this appear to be an **arid, poorly vegetated region** or a **humid, well–vegetated region**? (Circle one)

e. Based on your answer to letter d, is this limestone likely a strong, hard rock or a fragile, weak rock? (Circle one)

f. What types of mass wasting would this limestone likely result in? Slides, Slumps, Flows, or Falls? (Circle all that apply.)

g. Because limestone is formed from precipitation due to evaporation of water, what will happen if it is exposed to high quantities of water on a regular basis? _____

Figure 5.9 © ISTOCKPHOTO LP

h. If groundwater is constantly in contact with the lower layers of limestone, what problem does this result in? _____

_____

8. Of the four rocks you just looked at, conglomerate/breccia, sandstone, shale, and limestone, sandstone has a high permeability, meaning that water is able to freely flow through it. However, the conglomerate/breccia and limestone have a considerably lower permeability, and shale has the lowest, meaning that virtually no water can pass through it. What mass wasting problem does shale's low permeability cause for any rocks overlying it?

_____

_____

Just for Show

Go to the counter in the lab and take a quick look at some other materials sedimentary processes are capable of producing.

You have already looked at a photograph of limestone, which is one of several chemical sedimentary rocks created from dissolution and precipitation. While limestone results when calcite precipitates out of evaporating water, there are a number of different rocks that are formed due to other minerals precipitating from water evaporation. All chemical sedimentary rocks form thick, wide layers that extend the length of the water body, and grow higher as the water evaporates. These deposits are found primarily in arid environments, the leftovers from evaporating lakes and seas. Samples 1a, 1b, and 1c are all additional examples of rocks created from dissolution and precipitation.

Sample 1a: This rock is very descriptively called "Rock salt." Just pop the tip of your tongue on it and you will know why. This is what develops when the evaporating water supersaturates with respect to Na and Cl.

9.   What mineral makes up Rock Salt? _____

Sample 1b: This rock is known as alabaster and is made of the mineral gypsum. This is what develops when the evaporating water supersaturates with respect to Ca and SO_4.

10. What property does this rock have that would lead you to conclude it is made up of the mineral gypsum? _____

Sample 1c: This rock is known as chert and is made of the mineral quartz. This is what develops when the evaporating water supersaturates with respect to Si and O.

11. What two properties does this rock have that would lead you to conclude it is made up of the mineral quartz? _____ and _____

As we've discussed, some chemical rocks are produced with the aid of living organisms (plants or animals) and we refer to these as biochemical or organic chemical rocks. Rocks that are formed in this way require a combination of chemical and physical weathering process to create the rock. Samples 2a–2d are all biochemical rocks and Sample 3 is an organic rock. Note: Recall that biochemical rocks are composed of the hard parts of once living organisms (shells and skeletons), while organic rocks are made from the soft tissue of once living organisms. This is why we classify them differently.

Samples 2a, 2b, 2c, and 2d (Coquina, fossiliferous limestone, chalk, and micrite respectively). This group of rocks composes Earth's biochemical limestones. Essentially, organisms in the ocean waters will "ingest" the Ca and CO_3 ions released into the ocean by the rivers, which picked up the ions through chemical weathering of the continental crust during their journey over the continent. The ocean critters then use these ions to precipitate their calcitic (made of calcite) skeletons and shells. Once the organisms die, the calcitic skeletons and shells will sink to the bottom of the ocean where they will be physically weathered by other living organisms, ocean currents, and waves. Eventually, the broken-up pieces will become "lithified," just as in a clastic sedimentary rock forming a biochemical sedimentary rock. A coquina has loosely cemented, coarse shells and skeletal fragments; fossiliferous limestone is a mixture of visible fossil detritus scattered throughout a fine-grained matrix; and micrite is the equivalent of a calcitic mud—i.e., varied skeletal and shell fragments that have been weathered to microscopic particle sizes. Chalk is an earthy, powdery, light-colored rock—differentiated from other fine-grained biochemical limestones because it is composed of skeletal material from exclusively one group of organisms called coccolithophores.

12. Notice that all four of these rocks, as well as the crystalline limestone, are made of the mineral calcite. This being the case, what property would make all five of these rocks instantly identifiable?

_____

Sample 3 This is a sample of native coal. Coal forms when huge amounts of plant material are buried by mud quickly enough to prevent bacteria from decaying it. Over time, the pressure of overlying sediments squeeze the volatiles and water out of the vegetation, leaving behind a heap of carbon.

13. Pick up the sample and examine the texture and strength of the rock. Even if coal is found in thick layers, as it often is, would it add strength or weakness to the slope? _____

Explain: _____

_____

A few special cases

Sample 4 This rock is the product of a weathering process called "hydrolysis." This differs from chemical rocks made by dissolution and precipitation, in that water actively participates in the chemical reaction creating a new group of minerals known as "clay." Rocks formed from large amounts of clay are called claystones. There is a specific group of clays called "smectites" that behave almost like a sponge, in that they soak up water, which causes them to expand, and later releases the water, causing them to shrink.

14. Name a few ways in which this "shrink/swell" behavior can increase the frequency of landslides.

_____

_____

_____

Sample 5 This beautiful rock is called an "oolitic limestone." It forms when sand grains settle into a calcitic "ooze" at the bottom of the ocean. The calcite hardens around the sand grains in layers, creating "ooids" or calcite-coated spheres. When the ooids get lithified, an oolitic limestone is formed. As you can see, this rock is easily recognizable by the "bulls-eyes" in combination with the fact that the rock will _____, because it is made of calcite. (One of these days when I'm rich but not famous I'm going to have my bathroom re-done in oolitic limestone.)

Part C: Assessing the Landslide Hazard

Various studies have shown that the most damaging slope failures are closely related to human activities. Regulating these activities before land use takes place can substantially reduce loss. While no area is immune to slope failure, some areas are more likely to experience catastrophic landslides than others. What should you look for?

A. **Steep slopes:** Gravity is the force that causes downslope movement, so avoid steep slopes. The gentler the better.

B. **Weak bedrock:** Rocks that are poorly cemented or that have planes of weakness such as faults, foliation, fissility, and jointing are by nature unstable.

C. **Areas that lack vegetation:** Plant roots bind soil and stabilize it. Bare areas are much more likely to slide downhill catastrophically than vegetated areas. If there is a bare area on a vegetated slope, that was likely the result of a previous mass wasting event.

D. **Areas with heavy rainfall:** Heavy precipitation increases the pore pressure and separates the grains. It also liquidates the sediment, resulting in flows.

E. **Slope undercutting:** Landslides are particularly common along stream banks, highway road cuts, and sea coasts, as all the natural support has been removed.

F. **Earthquake-prone areas:** Shaking can destabilize slopes by decreasing the friction provided by grain-to-grain contacts. The result can be wholesale slope failure when a steep gradient is present or liquefaction if the area contains a substantial amount of mud or clay, regardless of the slope angle.

G. **Bending, misalignment, or tilting of objects on the slope:** Bending of trees, misalignment of fences and roads, and tilting of objects like gravestones and telephone poles are all indicators of destructive creep. If the mass wasting has progressed to solufluction, the land in question will have tongue-shaped lobes.

H. **Talus pile:** The presence of boulders or talus piles at the base of a cliff is demonstrating the area is prone to frequent mass wasting.

I. **Spoon shaped depressions**: Indicates slope is prone to slumping.

Because scientists and engineers understand the factors that make slopes unstable, prevention is an attainable goal in many regions. Exceptionally dangerous areas can be avoided, and useable areas can be stabilized by structures that include retaining walls, drainage ditches, subsurface piping systems, and various coverings, or through the use of such techniques as grading, planting trees, and cut-and-fill operations. Figure 5.10 shows a number of slope stabilization methods.

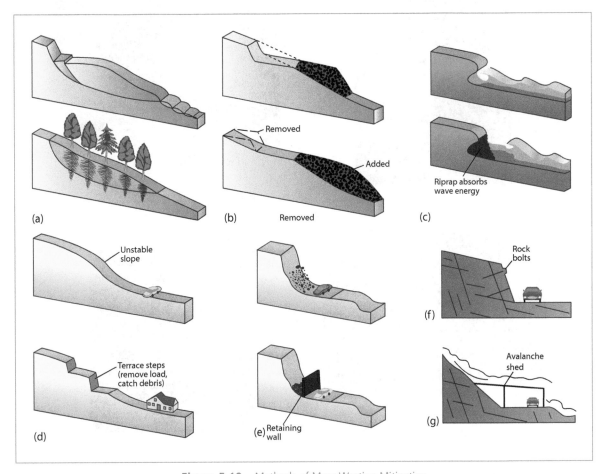

Figure 5.10 – Methods of Mass Wasting Mitigation

Below are a number of scenarios concerning landslide assessment and mitigation practices. Examine the photographs for each scenario, then answer the questions that follow.

Figure 5.11 © TOMASZ KURAN

Figure 5.14a © DARA JASUMANI

Figure 5.12 U.S. GEOLOGICAL SURVEY

Figure 5.14b © DARA JASUMANI

Figure 5.13 U.S. GEOLOGICAL SURVEY

Figure 5.14c © NIGEL CHADWICK

15. Examine figures 5.11–5.14 above. What mass wasting problem are these areas suffering from?

_____

16. What evidence do you see in each figure to support your answer in question 15?

Figure 5.11 _____

Figure 5.12 _____

Figure 5.13 _____

Figure 5.14a–5.14c _____

17. Examine Figure 5.15. What specific mass wasting problem is an accident waiting to happen in this location?

_____

Figure 5.15 © RICHARD LAW

18. What evidence do you see in Figure 5.15 that indicates large-scale mass wasting has happened here in the past?

_____

_____

_____

_____

19. Given that this is a lucrative tourist beach, closing the area is not an agreeable solution. If you were an engineer examining this area, what would you recommend be done to protect people along this beach?

_____

_____

_____

20. Examine Figures 5.16a–5.16f. What mass wasting problem is clearly prevalent in these locations?

_____

Figure 5.16a © RICHARD DORRELL

Figure 5.16b © COLIN VOSPER

Figure 5.16c © IAIN LEES

Figure 5.16d © NIGEL CHADWICK

Figure 5.16e © ANDY BEECROFT

Figure 5.16f © JIM CHAMPION

21. What two main features in these images led you to the answer you chose for #20? (There are two specific tell-tale signs of this type of mass wasting in each picture.)

- _____

- _____

22. Examine Figure 5.17. What type of mass wasting event occurred here? Slide, slump, flow or fall?

_____

23. What preventative measures could have been taken to reduce or even prevent the damage caused by this disaster?

_____

_____

Figure 5.17 © AP/FELIPE DANA

24. Examine figures 5.18a–5.18f. What type of mass wasting event occurred at these locations?

_____

Figure 5.18a © J. THOMAS

Figure 5.18b © RICHARD WEBB

Figure 5.18c © EVELYN SIMAK

Figure 5.18d U.S. GEOLOGICAL SURVEY

Figure 5.18e © BOB FORREST

Figure 5.18f © EIRIAN EVANS

25. What measures can be taken to reduce or even prevent this problem from occurring?

_____

_____

SOIL AND GROUND SUBSIDENCE

INTRODUCTION

Soil is the top layer of Earth's surface and is composed primarily of sediment (pebbles, sand, silt, mud, and clay particles) in combination with plant and animal debris. Soil is the material on which we grow our crops, build our cities, store our trash, and harvest our groundwater, among many other essential uses. As such, the stability of the soil in any given location is of pivotal importance for one reason or another.

Subsidence, the shrinking or settling of Earth's surface materials, has been attributed to surface compaction from withdrawal of fluids (oil, gas, and water), underground mining, decay of organic materials, and formation of sinkholes by collapse of bedrock in limestone karst regions. As such, subsidence is a hazard commonly associated with resource extraction. With increased need for resources extracted from the subsurface and increased human use of the surface, the problem of land subsidence can be expected to increase.

OBJECTIVE

The objective of this lab is threefold: 1) investigate the causes and hazards of land subsidence associated with fluid withdrawals; 2) examine case studies in which ground subsidence occurred as a result of water extraction and inappropriate use of organic soils; and 3) explore the causes of ground subsidence in limestone karst regions as well as the difficulties it poses for land-use planning.

© SOUTHWEST FLORIDA WATER MANAGEMENT DISTRICT

Part A: Land Subsidence Due to Water Extraction

Water, oil, and gas are all fluids that travel between the grains of Earth's subsurface materials. Yet rather than weakening the rock or sediment, the high fluid pressure they exert adequately supports the materials above, such that the grains are essentially locked in place (Figure 6.1a). If, however, the fluid is removed from overlying Earth materials, the support is reduced and subsidence will result (Figure 6.1b).

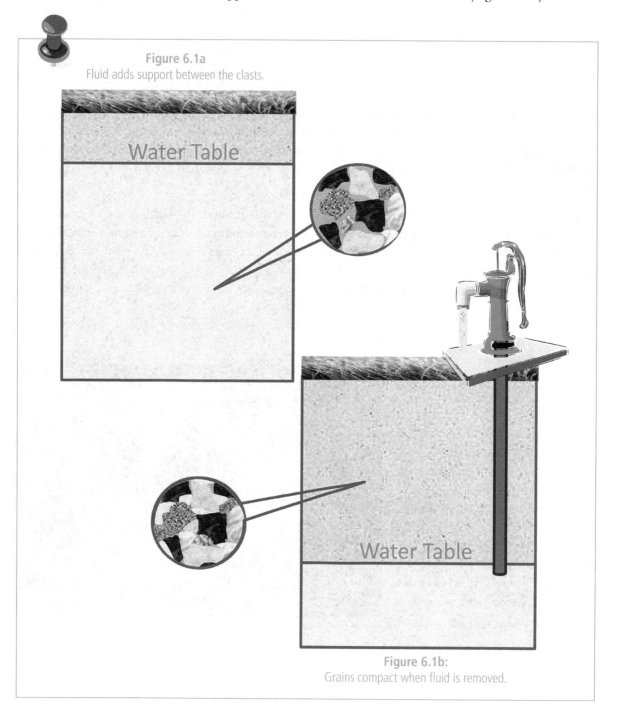

Figure 6.1a
Fluid adds support between the clasts.

Water Table

Water Table

Figure 6.1b:
Grains compact when fluid is removed.

Water is housed and transported through "aquifers,"—sediment or rock in which the spaces between grains are interconnected in a canal-like network (a property called permeability) and therefore able to transport water easily. Examples of aquifers include gravel, sand, soils, and fractured sandstone and limestone. Aquifers are divided into two basic types (see Figure 6.2):

A. **Unconfined Aquifer:** Where water can rise directly to the surface unimpeded; usually composed of "alluvium," the name given to sediments that have been deposited by streams, i.e., gravels and sand.

B. **Confined Aquifer:** Aquifer that is separated from the surface by an "aquitard,"—sediments or rocks that do *not* transport water easily and therefore retard the motion of water (e.g., mudstone or claystone). While aquifers are commonly made from alluvium, confining layers are often made of fine-grained silt, mud, and clay.

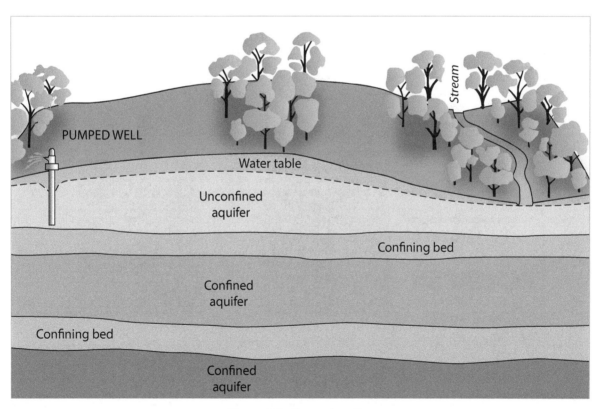

Figure 6.2 – Types of aquifers.

The causes of subsidence by water withdrawals are directly related to the compressibility of sediments and their contained water. When wells tapping an unconfined aquifer remove water from the sand and gravel of the aquifer, a widespread reduction of the water pressure in the aquifer causes compaction, thereby resulting in subsidence. It's important to note, however, that this decrease may be stopped or even slightly reversed if the fluid pressure is increased enough to cause the aquifer to return to its original shape. Thus, if the water is recharged at the same or similar rate it is being removed, subsidence is not a problem.

If a confined aquifer is present, a large part of the compaction, and therefore the surface subsidence, is due to the behavior of the clay-rich confining layers. Smectite clays are an unusual group of minerals that have the ability to swell when they absorb water and shrink when the water is removed. When water in a confined aquifer is lowered by pumping wells, water not only drains from the aquifer, but is also squeezed from the fine-grained confining beds. As such, when confining layers are present, the quantity of clay, the texture of the sediment, and the weight of the overlying load all affect the reduction in volume and therefore the amount of subsidence. Most ground subsidence occurs in confined aquifers, as shrinkage of the dried clay layers greatly contributes to the decrease in volume and thereby the settling of surface sediments (Figure 6.3).

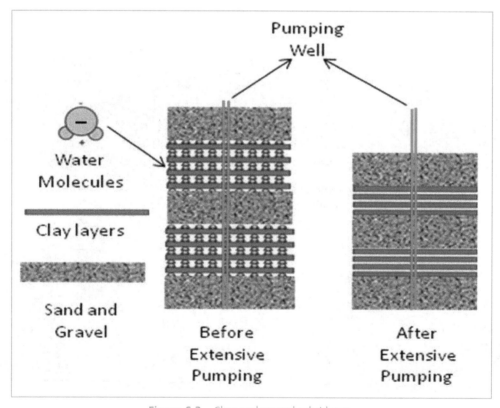

Figure 6.3 – Clays and ground subsidence.

1. The sediments in many major areas of subsidence consist of thick sand and gravel layers that are separated by layers of clay rich silt and mud. For each of the characteristics listed below, determine whether they are indicative of a sand/gravel aquifer or a clay rich silt and mud aquitard.

    - Aquifer   Sand/gravel or Clay Silt
    - Confining bed   Sand/gravel or Clay Silt
    - Low Compressability   Sand/gravel or Clay Silt
    - High Compressability   Sand/gravel or Clay Silt
    - Low Permeability   Sand/gravel or Clay Silt
    - High Permeability   Sand/gravel or Clay Silt
    - Fine Grained Texture   Sand/gravel or Clay Silt
    - Coarse Grained Texture   Sand/gravel or Clay Silt

2. Subsidence induced by groundwater withdrawal generally occurs in areas characterized by confined aquifer systems. How does a confined aquifer system differ from an unconfined aquifer system (i.e., what does a confined aquifer have that an unconfined one does not)?

    _____

3. Why would the presence of this extra component enlarge the amount of ground subsidence that occurs in a confined aquifer when water is over-drafted?

    _____
    _____
    _____
    _____

4. When the water pressure in the aquifer decreases, the water in the confining beds may move (**slowly** or **rapidly**) into the aquifer. The result is a _____ in thickness of the confining bed and possibly _____ of the land surface.

Part B: Ground Subsidence Case Studies

I. Santa Clara Valley, California

In the first half of the 20th century, the Santa Clara Valley was intensively cultivated and was nicknamed the "Garden of Eden" or "The Valley of Heart's Delight" for its myriad of fruit and vegetable orchards. As the population grew, so did the demand for food; between 1920 and 1960, an average of about 100,000 acre-feet per year of groundwater was used to irrigate crops.

Note: Hydrologists frequently use the term acre-foot to describe a volume of water. One acre-foot is the volume of water that will cover an area of one acre to a depth of one foot. The term is especially useful where large volumes of water are being described. One acre-foot is equivalent to 43,560 cubic feet or about 325,829 gallons!

The period from the end of World War II to approximately 1970 was a time of rapid population growth that was associated with the transition from an agriculturally based economy to an industrial and urban one. Nonagricultural use of groundwater began to increase substantially during the 1940s, and by 1960 total groundwater withdrawals approached 200,000 acre-feet per year. In 1964 the water level in the USGS monitoring well in downtown San Jose had fallen to a historic low, as groundwater was being used faster than it could be replenished.

Substantial land subsidence occurred in the northern Santa Clara Valley as a result of the massive groundwater overdrafts. In 1965 an integrated approach to groundwater recharge including surface water addition, storm flow water, and conservation of water during the rainy season led to substantial recovery of groundwater levels, and there has been little additional subsidence since about 1969.

The Santa Clara Valley was the first area in the US where land subsidence due to groundwater overdraft was recognized. It was also the first area where organized remedial action was undertaken and subsidence was effectively halted. The groundwater resource is still heavily used, but importation of surface water has reduced groundwater pumping and allowed an effective program of groundwater recharge that prevents groundwater levels from approaching the historic lows of the 1960s.

The valley is a large structural trough filled with over 1500 ft of alluvium, including a minority of sand and gravel aquifers, and the majority composed of fine-grained alluvium. Below a depth of 200 ft, the groundwater is confined by layers of clay, except near the margins. Initially, wells as far south as the town of Santa Clara flowed, but a decline in surface water pressure occurred due to pumping for agricultural purposes.

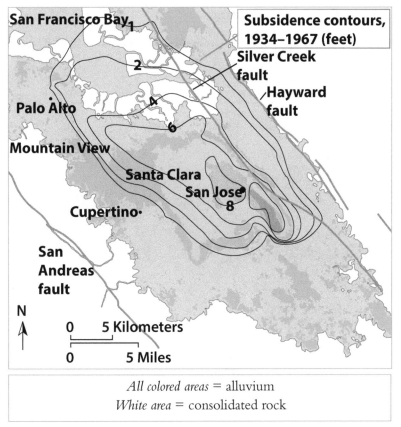

Figure 6.4 – Ground subsidence in the Santa Clara Valley, 1934–1967

Refer to Figure 6.4 to answer the following questions.

5. Which two locations showed the most subsidence in the Santa Clara Valley between the years 1934 and 1967? _____ and _____

6. How much ground subsidence occurred in these areas between 1934 and 1967? _____

7. How many feet per year did these areas subside during this period? _____ ft/yr

8. What geologic material was largely responsible for this subsidence? _____
    Explain why: _____
    _____

9. Assuming the ground subsidence problem was allowed to continue, would you expect the white area surrounding the alluvium in Figure 6.4 to eventually become affected? Why or why not?
    _____
    _____

Use Figure 6.5 to answer the following questions.

10. In 1965 the importation of water to the valley began. Some imported water was used to recharge the groundwater through stream channels.

   a. Between what years was the pumpage rate highest in this region? _____ and _____ .

   b. What was the pumpage rate during that time interval? _____ thousands of acre-feet

   c. Substantial water importation began around 1965. What was the pumpage rate by 1980? _____ thousands of acre feet

   d. Did the water level (aka "artesian head") rise or fall in response to the importation of water? _____

   e. By what percentage did the pumpage rate decrease from 1965–1980? _____ %

   f. By what percentage did the import rate increase from 1965–1980? _____ %

   g. Was there a substantial change in the amount of rainfall the region received between 1965 and 1980? **Yes** or **No**?

   h. Which of the following three factors was most influential in causing the water levels to increase from their low in the 1960s: **decreased pumpage**, **increased precipitation**, or **water imports**?

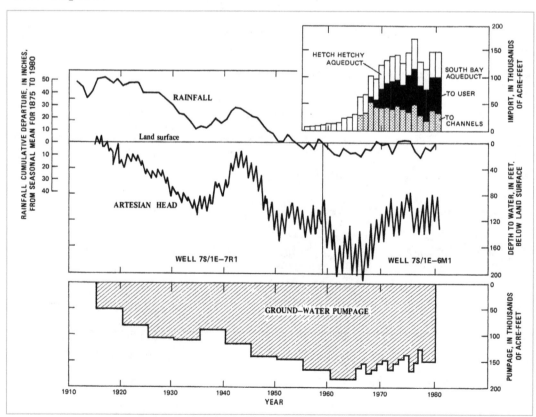

Figure 6.5 – Artesian head change in San Jose in response to rainfall, pumpage, and water imports. USGS

A second severely subsidence-plagued area in California is the San Joaquin Valley, also the result of extensive ground water mining for agriculture. While the withdrawal of water has allowed the San Joaquin Valley to become one of the world's most productive agricultural regions, it has also contributed to one of the single largest land-surface alterations attributed to mankind (see Figure 6.6).

11. Figure 6.7 is a graph of compaction changes in response to fluctuations in the depth of the water table in the San Joaquin Valley. Considering the data presented in this graph, answer the questions that follow.

   a. What is the relationship between groundwater levels and ground subsidence? _____

   _____

   _____

   Would subsidence resume in the valleys if imports of water were reduced and pumping of groundwater was increased? **Yes** or **No**?

   b. Examining Figure 6.7, notice that the "zero" line marks the area where there is no external compaction force. Does this mean the area sprang back to its pre-compacted thickness? **Yes** or **No**?

   Explain: _____

   _____

Figure 6.6 – Ground Subsidence in San Joaquin Valley. Signs on pole show approximate altitude of land surface in 1925, 1955, and 1977. Maximum subsidence in the San Joaquin Valley occurred near Mendota, California and was in excess of 28 ft.

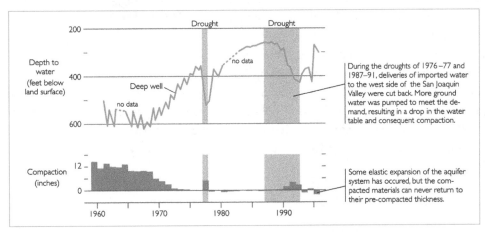

Figure 6.7 – Compaction changes in response to fluctuations in the depth of the water table in the San Joaquin Valley.

c. Based on your answer to letter **c** above, can the area ever be restored to its original condition? **Yes** or **No**?

II. Everglades, Florida

The Everglades ecosystem includes Lake Okeechobee and its tributary areas, as well as the roughly 40- to 50-mile-wide, 130-mile-long wetland mosaic that once extended continuously from Lake Okeechobee to the southern tip of the Florida peninsula at Florida Bay (Figure 6.8).

Since 1900, much of the Everglades has been drained for agriculture and urban development, so that today only 50 percent of the original wetlands remain. Water levels and patterns of water flow are largely controlled by an extensive system of levees and canals. The control system was constructed to achieve multiple objectives of flood control, land drainage, and water supply, but in drying out the naturally wet soils, it also induced severe land subsidence, which subsequently resulted in enormous fires during drought seasons and saltwater intrusion from the ocean. More recently, water-management policies have also begun to address issues related to ecosystem restoration. Extensive land subsidence that has been caused by drainage and oxidation of peat soils will greatly complicate ecosystem restoration and also threatens the future of agriculture in the Everglades.

Natural Flow Patterns

A thin rim of bedrock protects south Florida from the ocean. The limestone bedrock ridge that separates the Everglades from the Atlantic coast extends 20 feet or less above sea level. Under natural conditions all of southeast Florida, except for a 5- to 15-mile-wide strip along this bedrock ridge, was subject to annual floods. Much of the area was perennially inundated with fresh water. Water

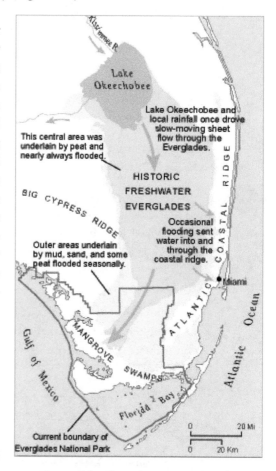

Figure 6.8 – Natural water movement in Everglades, Florida.

levels in Lake Okeechobee and local rainfall drove slow-moving sheet flow through the Everglades under topographic and hydraulic gradients of only about two inches per mile. Lake Okeechobee, which once overflowed its southern bank at water levels in the range of 20 to 21 feet above sea level, today is artificially maintained at about 13 to 16 feet above sea level by a dike system and canals to the Atlantic and Gulf coasts.

The high water levels and constant flooding in the vicinity of the Everglades posed numerous problems for farmers and urban developers looking to turn the swamp land into a "productive" region. Since 1913, a series of dikes, levees, and water-diversion canals have been implemented in an attempt to drain land needed for agriculture, to protect urban developments from flooding, and to route water to populated areas for human consumption (Figure 6.9). As the population grew, so did the need for water management, resulting in significant changes to natural surface flow patterns.

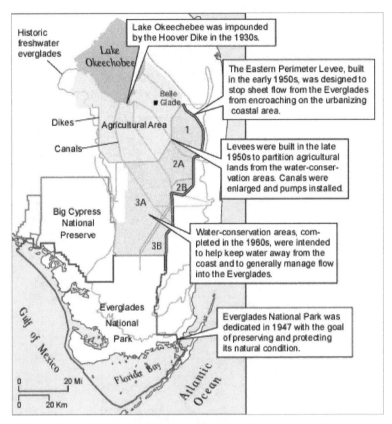

Figure 6.9 – Brief history of anthropogenic modification to the Everglades.

12. Figure 6.10 shows the water flow patterns in the Everglades region prior to anthropogenic interruption. Figure 6.11 depicts the same area after nearly a century of water flow modification for anthropogenic purposes. Compare figures 6.10 and 6.11 on the following page.

a. With respect to both the size of the area covered by flowing water and the flow velocity, what changes took place in:

- The northern Everglades just south of Lake Okeechobee
  Size of the flow area: **Increased** or **decreased**? By approximately what %? _____
  Flow Velocity change: **Significant, Moderate,** or **Low**?

- The central Everglades?
  Size of the flow area: **Increased** or **decreased**? By approximately what %? _____
  Flow Velocity change: **Significant, Moderate,** or **Low**?

- The western Everglades along the coast of Florida?
  Size of the flow area: **Increased** or **decreased**? By approximately what %? _____
  Flow Velocity change: **Significant, Moderate,** or **Low**?

- The southern Everglades around Shark Slough?
  Size of the flow area: **Increased** or **decreased**? By approximately what %? _____
  Flow Velocity change: **Significant, Moderate,** or **Low**?

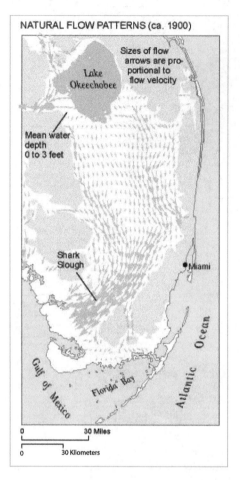

Figure 6.10 – Natural flow patterns in Everglades, Florida.

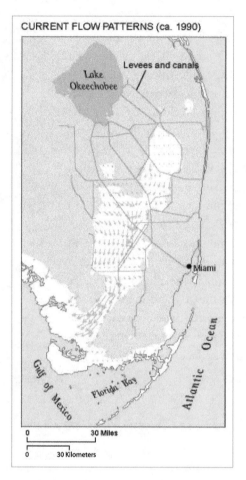

Figure 6.11 – Current flow patterns in Everglades, Florida.

b. Based on your assessment above, what negative impacts do you predict the following parts of the Everglades ecosystem will experience and *why*?

• Native Vegetation _____

_____

_____

• Native Wildlife _____

_____

_____

13. Water management has also significantly changed the natural vegetation patterns. Figure 6.12 shows the native vegetation distribution in the Everglades region prior to the construction of water control. Figure 6.13 depicts the same area after nearly a century of water-flow modification for anthropogenic purposes. Compare figures 6.12 and 6.13.

a. Notice that all Peripheral Wet Prairie, approximately 90% of the Swamp Forest and Saw Grass Plains, as well as half the Southern Marsh disappeared between 1900 and 1990. What is the reason for the disappearance of this vegetation? (*Hint:* What's in those spots now?)

_____

_____

b. The Cypress on the eastern side of the Everglades also completely disappeared along with about half of the Saw Grass Dominated Plains. Explain the reason for the disappearance of this vegetation: (*Hint:* What's in those spots now?)

_____

_____

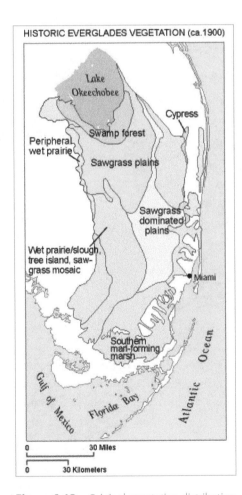

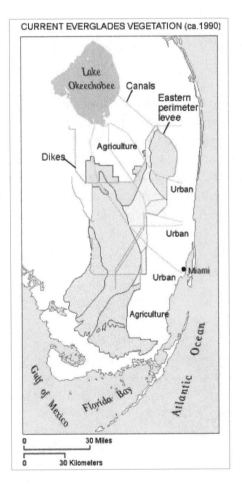

Figure 6.12 – Original vegetation distribution in the Everglades, Florida.

Figure 6.13 – Current vegetation distribution in the Everglades. Florida.

Drainage of the Everglades for agriculture and urbanization also resulted in significant ground subsidence, as organic material, such as peat and natural vegetative debris, are the primary sediment in wetlands, deltas, and other coastal environments. In addition to the fact that sediment grains compact and sink upon removal of the water, which in itself will cause subsidence problems, the organic material is subsequently exposed to corrosive oxygen in the atmosphere, resulting in the decomposition of the organic material, which thereby lowers the land surface. Finally, construction of dams, levees, and dikes to prevent flooding also compounds the problem, as the regular episodes of flooding bring a fresh sediment supply, which mitigates lowering of land height that results from natural, weight-induced sediment compaction. In conclusion, when left to nature alone, the peat was being replaced at a faster rate than it was being degraded, thereby maintaining an extremely slow (0.03 inches/yr) but steady accretion of peat to the land surface. However, modern drainage has disturbed the equilibrium such that the organic soils are being degraded at a faster rate than they can be replaced, resulting in a severe land subsidence problem.

Conventional surveying has always been extremely difficult in the Everglades due to the flexibility and instability of the surface material, as well the difficulties in accessing much of the terrain. Current best estimates suggest that there have been 3 to 9 feet of subsidence in the current Everglades agricultural area and that an equally large uncultivated area has experienced up to 3 feet of subsidence (Figure 6.14).

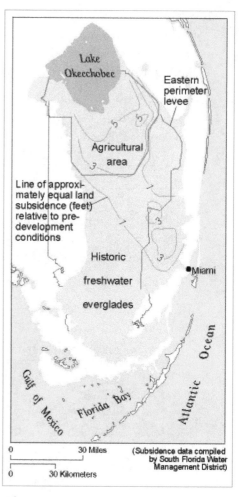

Figure 6.14 – Average land subsidence in the Everglades since 1913.

14. Based on Figure 6.14, which two areas have the highest rate of ground subsidence (i.e. 3–5 feet)? _____ and _____

Why? _____

_____

15. Consider the fact that in a near-sea-level wetlands system such as the Everglades, water flow is driven by less than 20 feet of total relief and the hydraulic gradient is only about 2 inches per mile. Based on that information, do you feel the loss of land surface in the highest areas of subsidence is significant? **Yes** or **No**?

Explain:_____

_____

_____

The Everglades agricultural area is now mainly devoted to sugarcane, with considerably smaller areas used for vegetables, sod grass, and rice. The value of all agricultural crops is currently about $750 million.

The crops require a thick, healthy layer of peat soil in order supply the necessary nutrients. Keeping this in mind, examine Figure 6.15:

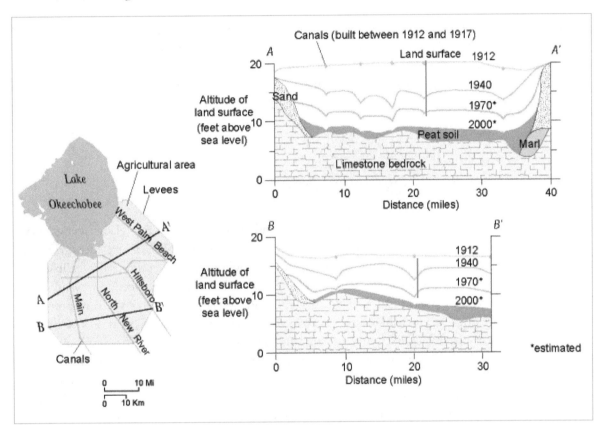

Figure 6.15 – Land subsidence in the agricultural area of the Everglades since 1912.

16. Using the data from 1912 and 1970 only in Figure 6.15, calculate the average rate of soil subsidence in the Everglades in the areas marked by the red lines.

_____ inches/yr on the A–A′ line

_____ inches/yr on the B–B′ line

Average subsidence in the area = _____ inch/yr

Note: This is actually considerably slower than in the Sacramento–San Joaquin Delta of California, the other major area of peat-oxidation subsidence in the United States, where average subsidence rates have been up to 3 inches per year. However, the pre-agricultural peat thickness was much greater in the Delta (up to 60 feet) than in the Everglades, where initial thicknesses were less than 12 feet.

17. According to the soil scientists, cultivation of the crops in this area would not be possible with less than one foot of soil. Knowing that the peat layer in its original, natural state began at a thickness of 12 feet and that the subsidence effects from the first canals began around 1915, use the average rate of subsidence that you calculated in #16 to determine in what year the peat thickness is expected to subside to approximately one foot._____

Although the extrapolation of peat thickness appears consistent with measurements made in 1969 (Johnson, 1974), 1978 (Shih and others, 1979), and 1988 (Smith, 1990), little land has yet been retired from sugarcane. One reason among several is that farmers have managed to successfully produce cane from only six inches of peat. However, it is clear that agriculture as currently practiced in the Everglades has a finite life expectancy, likely on the order of decades.

AND WHERE DO THE EVERGLADES STAND NOW?

Because of peat loss, it is believed that agriculture as currently practiced in the Everglades will gradually diminish over the next decades. However, subsidence makes true restoration of the Everglades agricultural area itself technically impossible, even in the event that it were politically and economically feasible. Land there that once had a mean elevation less than 20 feet above sea level has been reduced in elevation by an average of about 5 feet. Differential subsidence has significantly altered the slope of the land, precluding restoration of the natural, shallow sheet-flow patterns. If artificial water management and conveyance were now to cease, nature would likely reclaim the land as a lake, rather than the pre-development saw grass plains. With removal of the "sponge" of peat and native vegetation, the agricultural area has also lost most of its ability to naturally filter, dampen, and retard storm flows. Other strong impediments to restoration of the Everglades agricultural area include loss of the native seed bank, accumulations of agricultural chemicals in the soil, and the potential for invasion by aggressive exotic species.

In addition, even in the complete absence of agriculture in the Everglades, the existing pattern of urban development and land subsidence would prevent restoration of the natural flow system. Engineered water management and conveyance will be required indefinitely. Land subsidence over a large area south of Lake Okeechobee has created a significant trough within the natural north–south flow system, thereby preventing restoration of natural sheet-flow and vegetation patterns.

As such, "restoration" is perhaps a misnomer, as the focus of this restoration effort is on more natural management of the remaining 50 percent of the Everglades wetlands, not on regaining the 50 percent that has been converted to urban and agricultural use.

18. List 4 serious and likely permanent consequences humans will experience due to the draining of the wetlands in the Everglades:

   • _____

   • _____

   • _____

   • _____

III. New Orleans, Louisianna

Coastal Wetlands are vegetated, flat lying stretches of coast that flood with shallow water and are protected from violent wave activity making them the gentlest type of shoreline. They are comprised of organic soils—i.e., soil composed primarily of dead organic matter that is unable to decay due to the presence of shallow water that shields it from the oxygenated atmosphere. From a historical perspective, wetlands like swamps, bogs, and marshes, have mistakenly been considered insect infested wastelands, but in reality, are critical to the sustenance of the surrounding ecosystem, water resources, and the protection of coastal cities. Wetlands serve the multifold purpose of recharging groundwater supplies, removing pollution, soaking up floodwaters from storm surges, providing windbreakers during hurricanes and furnish countless fish and wildlife habitats. In fact, so many species use the wetlands to spawn that they account for 10–30% of marine productivity. Wetlands are also pivotal from an economic perspective because of their key role in fishing, hunting, agriculture and recreation. It was in this setting that New Orleans was built; and as the wetlands became further and further depleted, so did New Orleans's natural defenses against hurricane winds, storm surges, and saltwater intrusion.

The city of New Orleans resides on the gulf coast, nestled between two potential flooding threats—Lake Ponchartrain to the north, (which leads into the gulf), and the Mississippi river to the south (see Figure 6.16). The land between these two water bodies, in which New Orleans was built, is on average 10–12 ft below sea level (see Figure 6.17).

Figure 6.16 – Position of Modern Day New Orleans

Figure 6.17 – Cross Section (side view) of New Orleans

New Orleans was constructed within the Mississippi River floodplain. Before the region was colonized, the Mississippi River annually flooded the land now occupied by New Orleans, in the process depositing silt to create a vast expanse of wetlands. When the city was initially established in 1700, it was built on high ground adjacent to the Mississippi River—which was NOT annually flooded (Figure 6.18). This area would become what is now the French Quarter.

Due to the surrounding wetland environment, New Orleans couldn't grow until an engineer named Baldwin Wood designed a multifaceted system of pumps and drainage canals to drain the adjacent wetlands. Much of the same system is still in use today. With the flood-prone areas drained, New Orleans expanded outward from the French Quarter—all over the dried out wetlands. (See Figure 6.19).

Because the organic soils were no longer protected by water from the corrosive atmosphere, the organic material composted resulting in substantial ground subsidence. This caused the city to sink even further. In response, levees were built to protect the city from flooding.

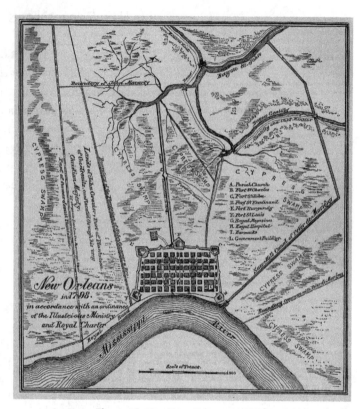

Figure 6.18 – New Orleans in 1798.

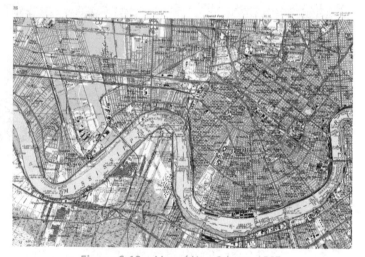

Figure 6.19 – Map of New Orleans, 1967

Figure 6.20 – Replenishing the wetlands

Before the levees were built, the Mississippi River was able to transport tons of soil and sediment to the coast. Each year, when the river underwent its seasonal spring flooding, fresh soil and sediment were added to the wetlands, replenishing bulk lost from compaction and delivering a fresh supply of nutrients to the wetlands biota (Figure 6.20). When levees were built, they prevented the river from undergoing its seasonal flooding cycle and the wetlands became starved of fresh sediment. As a result, the wetlands have been dematerializing at a rate of 20 mi^2 per year—which translates to nearly a football field an hour (see Figures 6.21a–6.21b). To complicate matters even further, while it was believed the levees and flood walls would prevent the city from flooding, there were no plans in place to mitigate the impending disaster in the event a levee or flood wall failed.

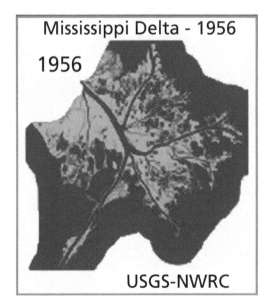

Figure 6.21a – Mississippi Delta 1956

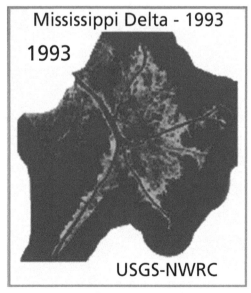

Figure 6.21b – Mississippi Delta 1993

19. Based on what you've learned above, list 4 or 5 reasons this city was a disaster waiting to happen.

 • _____

 • _____

 • _____

 • _____

 • _____

20. Suggest the consequences of wetland devastation on:

   a. Fresh water resources: _____

   _____

   _____

   b. Local ecology:

    • _____

    • _____

    • _____

    • _____

   c. Economy with respect to:

    • Fishing Industry _____

    _____

    • Agriculture _____

    _____

    • Tourism and recreation _____

    _____

Part C: Karst Topography

The term "karst" describes a landscape shaped by the dissolution of a layer or layers of soluble bedrock, usually carbonate rock such as limestone (Figure 6.22). The limestone dissolves because groundwater, after reacting with the rocks and minerals on the surface, becomes mildly acidic, permitting it to dissolve the calcite mineral crystals that make up the limestone.

Much of the drainage in karst areas occurs underground rather than as surface streams. Rainwater seeps into the ground among fractures in the bedrock (Figure 6.22), whereupon the acidic water dissolves the limestone around it. The cracks widen into narrow caves, which may eventually widen into huge cave galleries (Figure 6.23).

Figure 6.22 – Dissolution of carbonate rocks.

The systems of fractures and caves that typically develop in limestone are what make limestone a good aquifer.

Figure 6.23 – Underground karst cave network.

Karst regions are generally recognizable from the surface as they depict the following features:

- **Sinkholes:** Develop where the ceilings of cave galleries collapse, and lakes or ponds form wherever water fills the sinkholes (see Figure 6.24).

Figure 6.24 – Sinkholes

Figure 6.25 – Springs

Figure 6.26 – Disappearing Streams

- **Springs:** Places where water flows naturally discharge to the surface and flow from the ground.

- **Disappearing Streams:** Terminate abruptly by seeping into the ground.

21. Given that karst regions continually evolve, sinkholes that weren't present at the time buildings were constructed can and do open up later on, resulting in catastrophic damage to man-made structures in karst areas. Let's imagine you were building a log cabin bed-and-breakfast in the region depicted in Figure 6.27.

a. Which construction site (X, Y, or Z) is most likely to result in sinkhole-related problems?

_____

Why? _____

_____

b. Which construction site (X, Y, or Z) is least likely to result in sinkhole-related problems?

Why? _____

_____

_____

c. Being the sensible entrepreneur that you are, you decide the most responsible thing to do prior to building your bed-and-breakfast is to investigate the potential sites available and determine if there is a significant threat of sinkhole development in the future. How would you go about discovering if there is a significant sinkhole hazard in the locations you are considering building your bed-and-breakfast?

_____

_____

_____

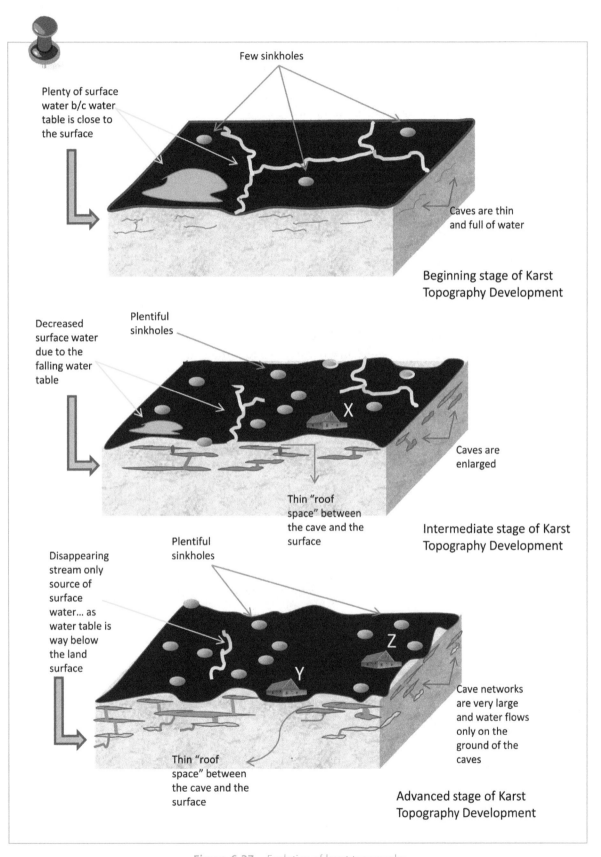

Few sinkholes

Plenty of surface water b/c water table is close to the surface

Caves are thin and full of water

Beginning stage of Karst Topography Development

Decreased surface water due to the falling water table

Plentiful sinkholes

X

Caves are enlarged

Thin "roof space" between the cave and the surface

Intermediate stage of Karst Topography Development

Disappearing stream only source of surface water... as water table is way below the land surface

Plentiful sinkholes

Z

Y

Cave networks are very large and water flows only on the ground of the caves

Thin "roof space" between the cave and the surface

Advanced stage of Karst Topography Development

Figure 6.27 – Evolution of karst topography.

22. Examine the Park City Quadrangle (Map 6-I). This map depicts a limestone karst region around Mammoth Cave, Kentucky. Sandstone overlies the limestone only in the area north of Park City.

    a. List two ways you can identify which part of the map has limestone at the surface and which part of the map has sandstone at the surface.

      • _____

      • _____

    b. Using a red pen, draw the boundary between the karst topography region and the area dominated by sandstone.

    c. Using a highlighter, find and trace the length of two disappearing streams.

    d. Take a red pen and circle two lakes that formed by flooding a sinkhole.

    e. You've decided to purchase a plot of land, where primarily sandstone is at the surface, for the purpose of establishing a farm. You select an area on one of the highest hills (~850 ft) located north of Park City, and make plans to drill a well into the limestone aquifer on your property for water to irrigate your crops. How deep would you have to drill your well (through the sandstone) to obtain water from the limestone aquifer? _____ ft

      Explain your reasoning: _____

      _____

      _____

      _____

    f. Why are there so few sinkholes developed in the SE part of this map area (even though limestone crops out there)? *Hint:* Drawing a quick sketch could be very useful here.

      _____

      _____

      _____

Map 6-I: Mammoth Cave, Kentucky

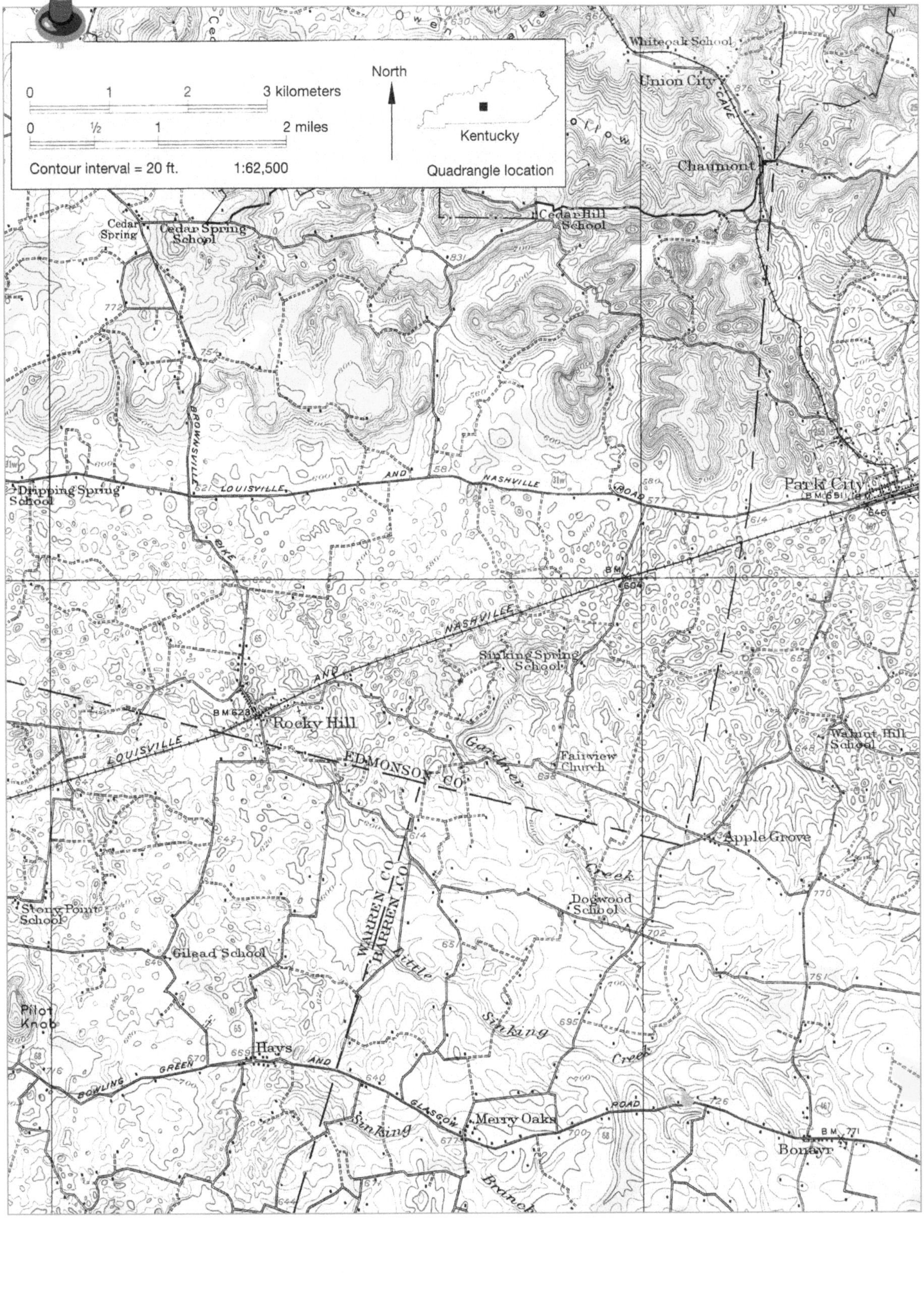

23. Map 6-II depicts a karst region in the northern part of Tampa, Florida called Sulfur Springs. The area is underlain by Floridian Limestone and has developed an extensive network of caves and sinkholes. Examine Map 6-II and answer the questions that follow.

a. Would you say the water table in this region is relatively close to the surface or considerably below the surface? _____ List two lines of evidence to support your answer.

   • _____

   • _____

b. Do you think the fracture and cave network in this region is minimal or extensive? _____
   On what did you base your conclusion? _____

   _____

   _____

c. Based on your answer to letter **b**, would you predict the groundwater flow velocity in this region to be **high** or **low**?

   Would the limestone in Sulfur Springs be considered an aquifer? **Yes** or **No**?

d. Population in Sulfur Springs has continued to rise in recent years and more people means more water use. Anthropogenic withdrawal of groundwater coupled with the natural evolution of a karst region can produce some devastating consequences. Describe what you expect to happen to each of the following features in this area, should this trend continue.

   • Surface streams, wetlands, and lakes: _____

   • Quantity of sinkholes: _____

   • Access to the groundwater (Hint: think about the height of the water table.) _____
     _____

e. Based on what you've determined about the fate of this area, list the threats that each of the following faces as a result of the predicted increase in groundwater overdraft.

   • Biota (plants and animals) associated with lakes, rivers, wetlands, and natural springs.

     _____

     _____

   • Stability of man-made structures such as homes, buildings, and bridges.

     _____

     _____

   • Availability of fresh water for urban consumption

     _____

     _____

Map 6-II: Sulphur Springs, Florida

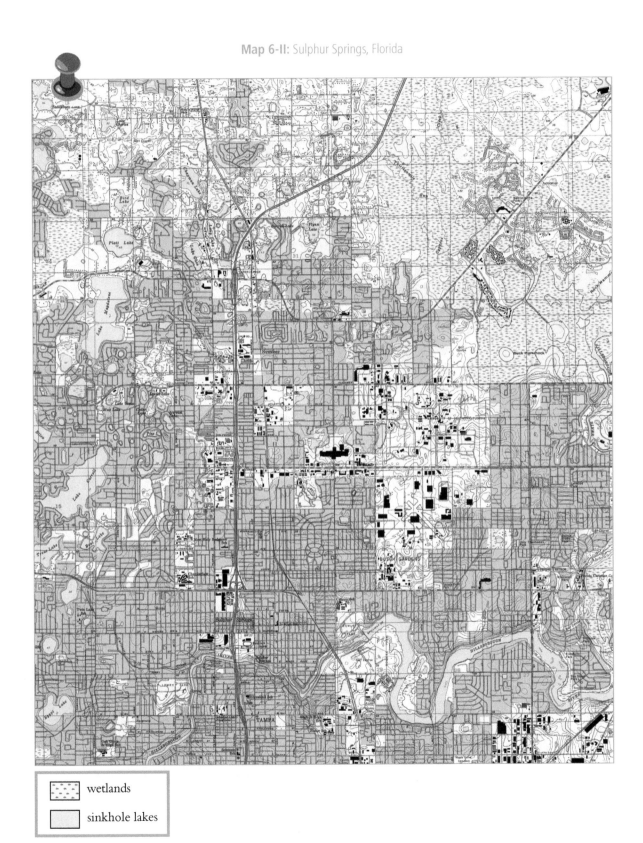

	wetlands
	sinkhole lakes

ROCKS IN MOTION AND THE HAZARDS THEY PRODUCE

INTRODUCTION:
PLATE MOTION – THE ROCK CYCLE AND THE CYCLE OF LIFE

Plate tectonics is a relatively new theory, developed in the 1960s, to explain why the continents appear to be "drifting" or changing positions over geologic time. Essentially it was discovered that Earth's lithosphere is broken up into several pieces called "lithospheric plates," each of which is constantly in motion on the slushy asthenosphere. Any processes associated with the creation, movement, and/or destruction of these plates is collectively called "plate tectonics."

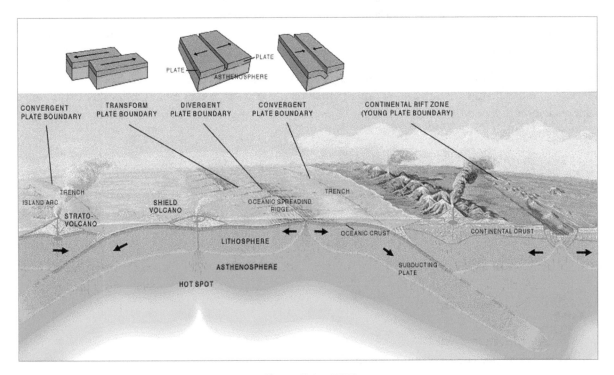

Figure 7.1 – USGS

According to the developing plate tectonics model (Figure 7.1), the continents are part of rigid lithospheric plates that move about relative to one another. Plates are generated along divergent boundaries, such as mid-ocean ridges, where two plates move away from each other. Magma then rises up between the two plates, fills in the gap, and cools to create new oceanic crust (basalt) on the edges of both plates. Plates are destroyed along convergent boundaries, where the edge of an oceanic plate subducts beneath an oceanic or continental plate, causing the subducting plate to be pulled back into internal Earth. When an oceanic plate subducts beneath another plate, a series of heat-induced reactions on the slab take place that result in the production of magma, which floats up and cools to form new continental crust (grano-diorite). Over time, wind, water, and other surface elements will erode the continental crust and transport the sediment across the continent, where it can be used for nutrients and resources. Those same processes will ultimately bring the spent material to the deep ocean, where it too will be subducted. If there is a continent at the boundary of *both* plates over a subduction zone, the two continents will collide and form a mountain range due to the crumpling and suturing of the two landmasses, as continental crust is too buoyant to be subducted. Plates may also slide past one another along transform fault boundaries, where plates are neither formed nor destroyed. Earth's size remains constant, because there are processes that simultaneously form and destroy lithospheric crust, thereby keeping production and destruction in equilibrium.

The tectonic cycle is in many ways the backbone of Earth's ability to support biology. It is through the cycles described above that fresh material is regularly brought to Earth's surface for biological consumption, and old material is returned to Earth's interior for a fresh supply of nutrients. If this process ever stopped, biology would soon become extinct, as there would be no supply of nutrients with which to grow crops and no supply of material to provide us with shelter and the myriad of items we depend on for day-to-day survival.

However, as natural and necessary as this process is, it too comes at a price. Earthquakes, volcanic eruptions, landslides, and tsunamis are just a few of the natural hazards plate tectonics brings to local environments. Understanding why and how these hazards come about is a pivotal part in being able to adapt to the changes they produce.

OBJECTIVE

The objective of this lab is threefold: 1) acquaint you with the fundamental characteristics of plate tectonics; 2) identify the hazards each part of the cycle produces; and 3) learn how the rock record preserves the history of these hazards and, when read properly, can in turn warn us of upcoming dangers.

Because the mitigation practices for the four major hazards discussed above have already been or will be addressed, hazard prevention is omitted from the current assignment.

Part A: Faulting

As discussed above, plates can interact with each other in one of three ways. They can collide, pull apart, or grind past each other. As such, there are three kinds of directed force (stress) that can be applied to the solid mass of rock (Figures 7.2 and 7.3 below). Compression compacts a block of rock and squeezes it into less space. Tension pulls a block of rock apart and increases its length. Shear smears a block of rock from side to side and will eventually tear it apart into two blocks of rock that slide past each other.

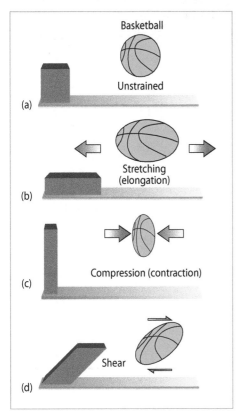

Figure 7.2

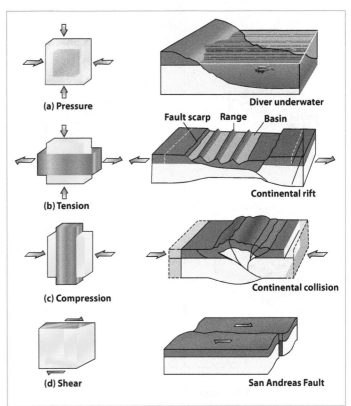

Figure 7.3

1. Analyze the three block diagrams in Table 7.A. Complete the three blank columns in the diagram to infer what kind of stress and strain is associated with each kind of faulting and plate boundary.

Table 7.A – Fault types and motion.

Fault Type	Shape Change? (Shortened, Elongated, or Neither)	Type of Stress? (Tension, Shear, or Compression)	Plate Boundary Type? (Convergent, Transform, or Divergent)
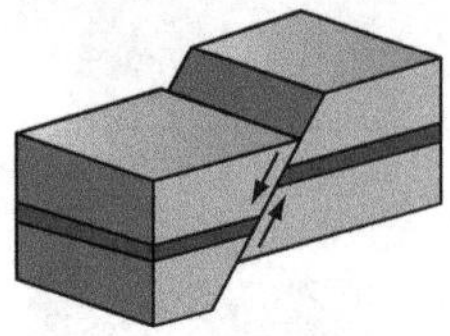 **Normal**			
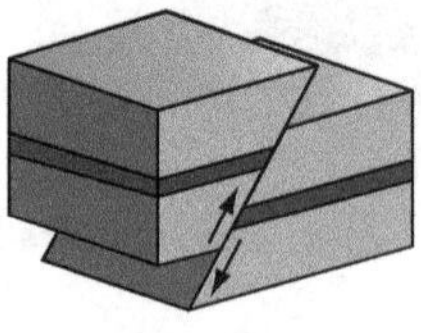 **Reverse**			
Left-lateral Right-lateral  **Strike-slip faults**			

Part B: Rocks subjected to heat and pressure

As discussed in the previous exercise, major rock masses will collide, pull apart, or grind past each other as a consequence of tectonic movement—generating enormous stress on the rocks. If rocks on the cold, rigid surface were exposed to pressures of this magnitude, they would crack apart in response to the stress. However, rocks that encounter these stresses deeper in the Earth's interior behave very differently, as these rocks are also heated and are therefore able to deform without rupture. Metamorphism is a process that induces changes in a rock's texture and composition without shattering or melting the original material. Rocks that undergo textural and mineralogical changes due to experiencing temperature and pressure conditions different from those in their environment of formation are termed metamorphic rocks. Metamorphic rocks are the history books for the stressed and deformed rock zones associated with tectonic environments and contain many layers of information from what they were originally to the process that took place, which resulted in their alteration.

Metamorphic rocks are generated from the combined effects of heat and pressure—both of which increase with depth. While temperature increases due to processes taking place far inside the Earth, pressure increases with depth due to the increasing thickness of overlying rock. All rocks that are brought from shallow to greater depths will experience "confining pressure"—a unique type of pressure in which the rock is subjected to the same degree of force from all directions. Thus, when confining pressure is applied, because the rock is being "squeezed" equally in all directions, the rock will become denser and decrease in volume, but distortion to the shape and/or orientation does not result.

In contrast to the confining pressure that all metamorphic rocks are exposed to, isolated circumstances may expose certain rocks to differential stress—a situation in which the forces are not equally applied from all directions—consequentially stressing the rock more from one direction then from another. The rock will respond by stretching, compressing, or shearing—thus resulting in distortion to the shape of the rock and/or the minerals composing it.

Since all metamorphic rocks are exposed to elevated temperatures and confining pressure, it is the presence or absence of evidence for differential stress that provides the first tier of classification for metamorphic rocks—foliated or nonfoliated.

Foliated Rocks

Foliation (see Fig. 7.4) refers to a metamorphic texture in which flat or elongate minerals are preferentially aligned in response to exposure to high temperatures and differential stress in a high pressure environment. Rocks that display foliation are said to have undergone "regional metamorphism"- as the most common place for foliated rocks to form is in transform or reverse fault zones (both of which extend hundreds and sometimes thousands of miles)—where the shearing in the former, or compression in the latter, generate tremendous differential stress.

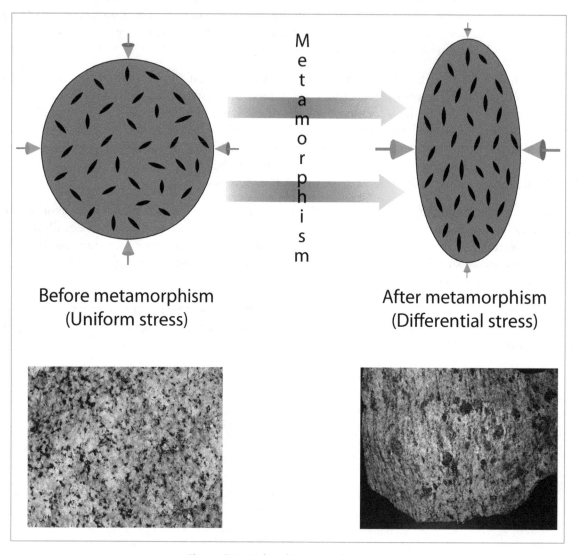

Figure 7.4 – Foliated Metamorphic Rocks

Non-Foliated Rocks

If the rock has been exposed to high temperatures and pressures—but no differential stress—the rock will not show any foliation. Instead, the textural identifier for these rocks is fused-together grains, which are created by a process called "recrystallization," in which the atoms of the original, individual grains, break their bonds and then re-bond as one entity, forming a mosaic-type texture (see Figure 7.5).

These rocks are usually formed when a magma intrusion makes it way up to the surface and "cooks" the surrounding rock. Metamorphic rocks that do not display foliation are said to have undergone "contact metamorphism"—as only the rocks that came in direct contact with the heat from the rising plume are metamorphosed. As such, unlike regional metamorphism, contact metamorphism affects only a localized area.

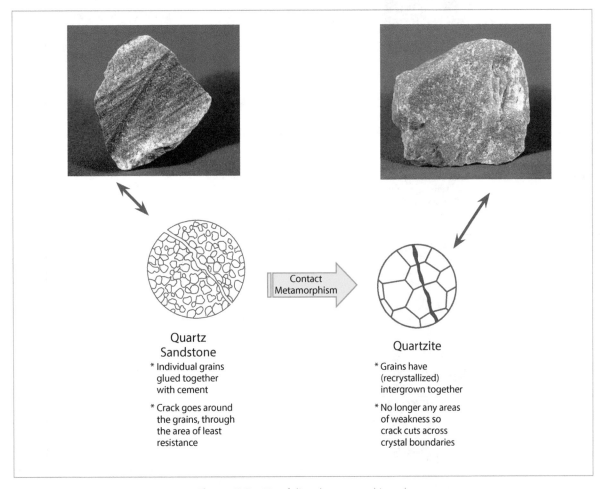

Quartz
Sandstone

Contact
Metamorphism

Quartzite

* Individual grains
glued together
with cement

* Crack goes around
the grains, through
the area of least
resistance

* Grains have
(recrystallized)
intergrown together

* No longer any areas
of weakness so
crack cuts across
crystal boundaries

Figure 7.5 – Non-foliated metamorphic rocks

Types of Metamorphic Rocks

	Rock	Hand Sample	Type of Foliation	Degree of Metamorphism	Description	Protolith (Parent Rock)
Foliated	Slate		Slaty cleavage	very low	Alignment of clay particles; fine grained, very slight buckling and pinching	Shale
	Phyllite		Phyllitic luster	low	Alignment of small micas; more coarse than slate, but still fine grained; clearly visible buckling and pinching	Shale
	Schist		Schistosity	moderate	Alignment of large micas; medium grained; Buckling, swirling, and pinching	Shale
	Gneiss		Gneissic	high	Separation of mafic and felsic minerals (i.e. gneissic banding). Coarse grained	Granite, diorite, gabbro
	Amphibolite		--	moderate	Shiny, black, bladed amphibole	Basalt, gabbro
Foliated or Non-foliated	Marble		--	--	Coarse, interlocking, fused together grains. Fizzes with HCL	Limestone
	Quartzite		--	--	Fine, interlocking, fused together grains. Scratches glass.	Sandstone
	Serpentinite		--	--	Contains asbestos	Basalt
	Hornfels		--	--	Cooked fine grained rock	Shale, Siltstone, Basalt
	Metaconglomerate/ Metabreccia	Metaconglomerate	--	--	Large clasts that have been fused together	Conglomerate, Breccia

Part C: Tying It All Together—Tectonic Environments, Associated Stresses and Hazards, and the Metamorphic Rocks Each Produces

Exercise 1: Spreading Center

Seafloor spreading takes place at an underwater network of fractures called "midocean ridges" (i.e., spreading centers—see Figure 7.6) that are generated at the interface of the cool, rigid crust and the hot mantle. As cracks open up in the oceanic lithosphere, the asthenosphere decompresses and melts. The magma then rises up to fill in the crack—where it cools to create new oceanic crust. This process is then repeated indefinitely.

2. What kind of motion is taking place here? **Divergent, Convergent,** or **Strike-slip.**

3. What kind of stress is this fault undergoing? **Tension, Compression,** or **Shear.**

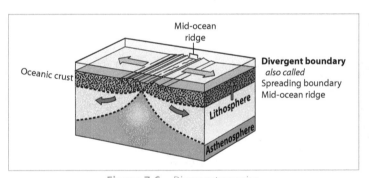

Figure 7.6 – Divergent margins

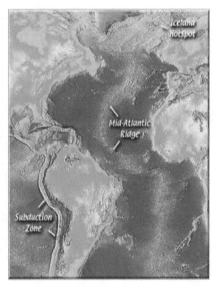

The Mid-Atlantic Ridge, which splits nearly the entire Atlantic Ocean north to south, is probably the best-known and most-studied example of a divergent-plate boundary.

4. What primary type of fault would you expect at this location? **Normal, Reverse,** or **Transform.**

5. Look again at Figure 7.1, located at the beginning of the lab. Observe that the fissures in a mid-ocean ridge (spreading center) do not form a straight line, but are actually composed of segments that are offset from each other. What additional type of fault does this setup create in spreading center regions? _____. What kind of stress would you expect to find there? _____

6. What hazards are spreading centers responsible for producing? **Volcanism, Earthquakes,** or **both**?

7. Are any of the hazards generated by a spreading center a serious threat to humans? **Yes** or **No**? Explain. _____

_____

Exercise 2: Subduction Zones

Plates are destroyed along convergent boundaries, where the edge of one oceanic plate subducts beneath another plate, resulting in the subducting slab being pulled back into Earth's interior. As the oceanic plate descends into the hot mantle, water released from the subducting slab floats up and lowers the melting temperature of the upper mantle, resulting in the production of magma, which in turn migrates upward and produces volcanoes.

When an oceanic plate descends beneath another oceanic plate, the rising magma will form an "island arc," a chain of volcanic islands that parallel the convergent plate boundary. See Figure 7.7.

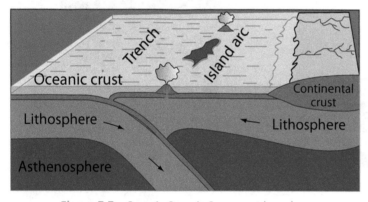

Figure 7.7 – Oceanic-Oceanic Convergent boundary.

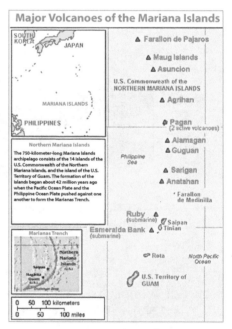

Oceanic-Oceanic Plate Boundary
Example: Mariana Islands

8. Examine Sample A. Sample A is a rock that forms when basalt gets compressed in a high temperature, high pressure environment. Identify sample A.

_____

9. Looking at Figure 7.7, this would be the rock that forms between the black a rows on the subducting slab. Notice the grains are all aligned in this rock. This is an example of foliation. Why are the grains arranged this way?

_____

_____

_____

When an oceanic plate descends beneath a continental plate, the rising magma will float up and add to the existing continental crust, creating a chain of composite volcanoes on the overriding continent (see Figure 7.8).

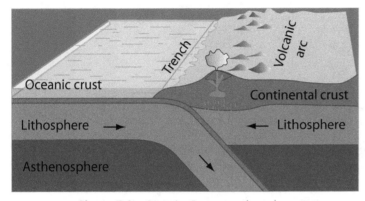

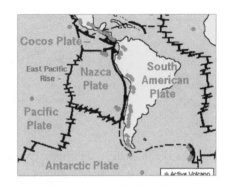

Figure 7.8 – Oceanic–Convergent boundary

Oceanic-Continental Plate Boundary
Example: Andes Mountain range.

As oceanic crust subducts beneath a continent, sediments that have been transported off the continent, as well as marine sediments scraped off the down-going slab, become wedged between the two colliding plates, forming a delta-shaped feature called an accretionary wedge/prism (see Figure 7.9). The wedge undergoes metamorphism due to higher temperatures and pressures associated with burial and compression resulting from the two colliding plates. As such, the accretionary wedge experiences a phenomenon called "tectonic foliation" (see Figure 7.10) in which the platy and elongate grains realign, and in some cases equant grains deform (see Figure 7.11) in addition to metamorphic changes in mineralogy.

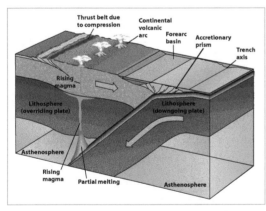

Figure 7.9 – Position of accretionary wedge.

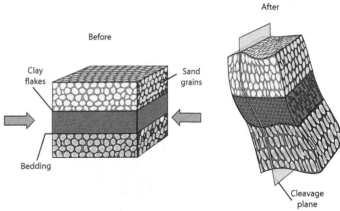

Figure 7.10 – Tectonic Foliation.

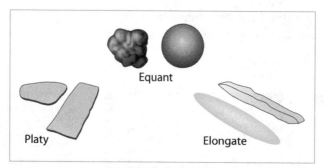

Figure 7.11 – Equant grains do not display foliation unless deformed, as there is no preferred orientation

Sample B is a shale, a sedimentary rock formed when mud-sized particles become lithified. When shale gets exposed to increased temperatures and pressures, as it does in an accretionary wedge, the rock increases in density. The new, denser rock has very fine ribbons running through it—a very subtle indicator that the rock was compressed in a high-temperature, high-pressure environment. As this rock gets buried further and subjected to even higher temperature and pressure conditions, muscovite, the primary mineral in the mud that composed the original shale, will become coarser and start to be visible to the naked eye. Eventually, the continually increasing stress will force the muscovite to develop clear, coarse crystals, and metamorphic minerals, such as garnet, begin to form. After this stage, any additional stress added to the rock will result in melting, and the rock will turn to magma, and return to the surface through volcanism in due time. As such, shale goes through three distinct metamorphic stages due to its being buried and compressed in an accretionary wedge environment.

Samples C, D, and E each represent one of the three metamorphic stages shale transforms to while stressed in an accretionary wedge environment.

10.

    a.   Which sample represents the first metamorphic stage? Sample C, Sample D, or Sample E? (Circle One)

    b.   What is your evidence? _____

    c.   What is the name of this rock? _____

11.

    a.   Which sample represents the second metamorphic stage? Sample C, Sample D, or Sample E? (Circle One)

    b.   What is your evidence? _____

    c.   What is the name of this rock? _____

12.

    a.   Which sample represents the third metamorphic stage? Sample C, Sample D, or Sample E? (Circle One)

    b.   What is your evidence? _____

    c.   What is the name of this rock? _____

Observe Samples F and G. Sample F is basalt, the same rock that makes up the oceanic crust, and sample G is the rock that forms when basalt is altered by hydrothermal fluids that are circulating throughout the subducting slab. Over time, slivers of metamorphosed oceanic crust are sheared off the down-going slab during subduction and become a part of the accretionary wedge. Notice the threadlike mineral you see in sample G. This mineral is asbestos, the very same material that is used to insulate homes and buildings, cushion brake pads, and make asbestos fabric, which is used as a fire retardant. Pay close attention to the crystal structure of the asbestos crystals. Notice they are shaped like tiny darts (see Figures 7.12a and 7.12b). It is this property that makes *airborne* asbestos dangerous in *large* quantities. Essentially, when these particles are freely floating in the atmosphere, they can be inhaled and will then treat the lungs like a dartboard, resulting in lung cancer after significant exposure. Have no fear here though. As you can see, these particles are not loose in the atmosphere, but are attached to the rock, there is an extremely small quantity, and your exposure is infinitesimal—so enjoy the view in a safe environment, as you likely will not get to see this again!

Figure 7.12a – Asbestos fibers © EURICO ZIMBRES

Figure 7.12b – Scanning electron microscope image USGS

13. Use the Types of Metamorphic Rocks chart to identify Sample G. What is the name of this rock?

Now refer back to Figure 7.9. Notice that the magma generated in the subduction zone from wet melting of the upper mantle must pass through the continental crust, and several layers of sedimentary rock before exiting from the volcano. In this part of the subduction zone, the shallow rocks are subjected to extremely high temperature, but no differential stress and significantly lower confining pressure in comparison to the accretionary wedge environment. For this reason, rocks that are affected by the upwelling magma do not show foliation, but instead display a texture called "fused-together grains," where the high heat instigated a reaction that fused together what were individual grains or crystals into an interlocking texture.

Compare Samples H and I. Sample H is a conglomerate, the same sedimentary rock you have kicked around outside thousands of times. Sample I is a rock formed when sedimentary conglomerate gets cooked by magma upwelling from a subducting slab. Look closely at Sample H's texture and notice how the grains have been "fused" together. Compare that texture to the way the sediment is "glued" together in the original conglomerate.

14. Identify Sample I: _____

15. Which do you think is stronger, i.e., less susceptible to breaking: the sedimentary conglomerate or Sample I? _____

    Explain:_____
    _____
    _____

When a rock is exposed to differential stress, it will only display the property of foliation if the grains are thin in one direction and elongate in another (i.e., flat, columnar, or dart-shaped). If, however, the rock is made from *undeformed* sand grains or calcite crystals, which have the same distance to all sides from the center of the grain, the rock might not appear foliated, even if it has been exposed to high heat and differential stress (refer back to Figure 7.11). Sample K is a quartzite, the rock formed when Sample J, a sandstone, is metamorphosed; Sample M is a marble, the rock formed when limestone, Sample L, is metamorphosed. Quartzite and marble can be found in environments of high temperature in conjunction with confining pressure, like the area in a subduction zone where magma rises to the surface, or one of high temperature and differential stress like the accretionary wedge. Compare the parent rocks of sandstone and limestone, with their metamorphosed counterparts quartzite and marble. Notice that both the metamorphic rocks have a texture called fused-together grains, but no foliation. Because these rocks have such similar textures, it is often difficult to tell them apart with the naked eye.

16. What two diagnostic tools could you use to immediately distinguish these two samples? _____ and _____

17. Sample N is a rock formed when mudstone is exposed to contact metamorphism.

    d.   Name this rock _____

    e.   Is this rock **foliated or non-foliated?** (Circle one.)

18. Does this make sense with respect to how it was formed? Yes or No? Explain. _____
    _____
    _____
    _____

Exercise 3: Suture Zones

When a continental plate collides with another continental plate (see Figure 7.13), no subduction occurs, as the continental crust is too buoyant and is unable to sink. In this case, the two continents will "suture," forming a mountain range. As such, the mountain range is partially created from continental crust and overlying sedimentary layers that buckled due to the collision, and continues to grow higher as one plate plows into another.

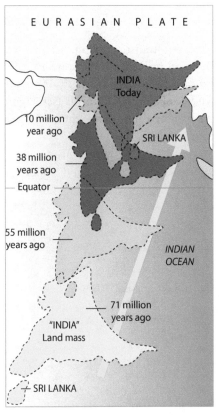

Continental-Continental Convergent Boundary Example: Himalayas where India collided with Asia.

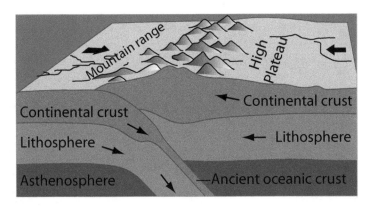

Figure 7.13 – Continental–Continental Convergent boundary.

The differential stress that continental crust undergoes due to mountain building is astronomical, and will cause the elongate and platy grains in the granite comprising the continental crust to rotate such that their longer side is aligned in the direction of least stress. Samples O and P are rocks that formed from such a process.

19. What is the name of these rock samples? _____

20. Do you see any resemblance between samples O and P and the granite (samples Q and R) that forms the continental crust? Yes or No? If yes, what? _____

    What's different? _____

21. What kind of motion is taking place in all three of these collision zones? Tensional, Compression, or Shear.

22. What type of fault would you expect due to a plate collision? Normal, Reverse, or Strike-slip.

23. What hazards are these processes responsible for producing? Volcanism, Earthquakes, or both?

24. Are any of the hazards you circled in #23 a serious threat to humans? Yes or No? Explain: _____
    _____
    _____
    _____

Exercise 4: Transform faults

Although transform faults are primarily located throughout spreading centers, where they are relatively harmless to humans, there are a few places on Earth where these faults develop as a result of entire plates grinding past each other in opposite directions. A classic example of this type of movement is the San Andreas Fault in California (see Figures 7.14a and 7.14b). If a continent is spanning across two plates that are grinding past each other, over geologic time, the continent will be ripped apart, each plate claiming its own piece.

25. What kind of motion is taking place here? **Divergent, Convergent, or Transform.**

26. What kind of stress is this fault under? **Tension, Compression, or Shear.**

27. What type of fault would you expect at this location? **Normal, Reverse, or Strike-slip.**

28. What hazards are these processes responsible for producing? **Volcanism, Earthquakes, or both?** (Circle all that apply)

29. Are any of the hazards you circled in #28 a serious threat to humans? **Yes or No?** Explain: _____

_____

_____

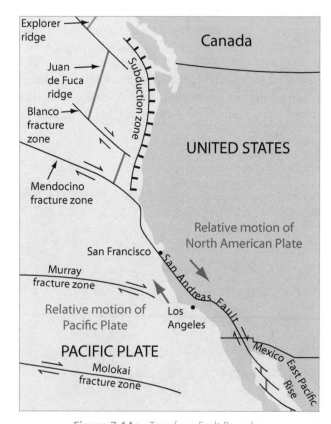

Figure 7.14a –Transform Fault Boundary.

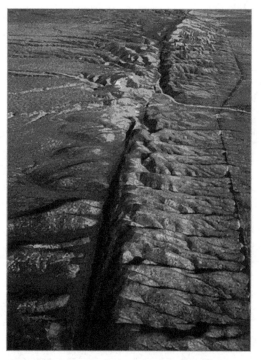

Figure 7.14b – Surface view of the San Andreas Fault.

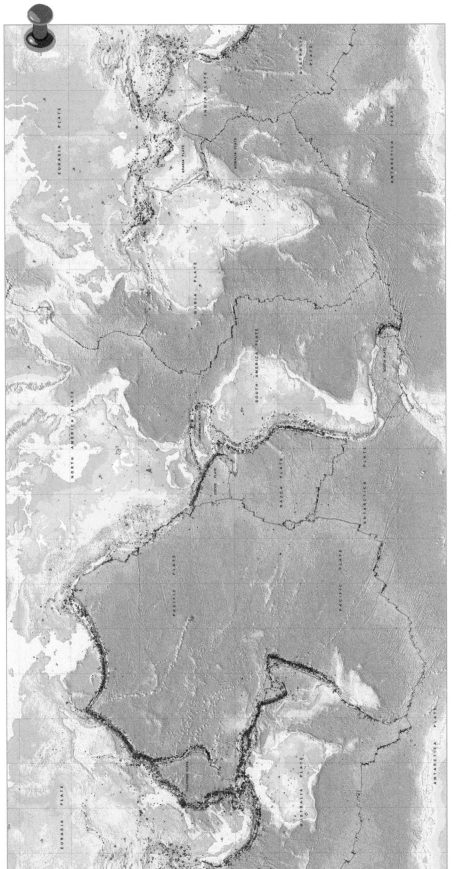

TOPOGRAPHY: DEPTH AND ELEVATION IN METERS

Sea-floor data from Scripps Institution of Oceanography Web site "Satellite Geodesy" (http://topex.ucsd.edu/marine_topo/mar_topo.html), accessed June 11, 1997; land-elevation data from U.S. Geological Survey (1997). Most lakes are shown in pale blue, with depth not implied. For the world's largest inland sea (Caspian) and deepest lake (Baikal), however, actual lake floor topography is shown using the same color scale as for the rest of the map.

Figure 7.15 – Relationship between plate boundaries, earthquakes, and volcanoes.

Volcanoes—Data from Global Volcanism Program, Smithsonian Institution, Washington, D.C.; accessed at http://www.volcano.si.edu/world/summary.cfm , March 16, 2005

▲ Erupted A.D. 1900 through 2003

△ Erupted A.D. 1 through 1899

△ Erupted in Holocene time (past 10,000 years), but no known eruptions since A.D. 1

△ Uncertain Holocene activity and fumarolic activity

Impact Craters—Data from University of New Brunswick, Planetary and Space Science Centre, Earth Impact Database; accessed at http://www.unb.ca/passc/ImpactDatabase/, October 23, 2003 (also see Grieve, 1998). Geologic age span: 50 years to 2,400 million years. Crater diameter indicated below

* <10 km

✳ 10 to 70 km

○ >70 km (shown at actual map scale)

Notable Events—Numbers next to a few symbols—of many thousands shown—denote especially noteworthy events, keyed to correspondingly numbered entries in tables found on the back of the map. These numbered events have produced devastating natural disasters, advanced scientific understanding, or piqued popular interest. They remind us that the map's small symbols may represent large and geologically significant events

19 Volcanoes

9 Earthquakes

23 Impact craters

Earthquakes—Data from Engdahl and Villaseñor (2002). From 1900 through 1963, the data are complete for all earthquakes ≥6.5 magnitude; from 1964 through 1999, the data are complete for all earthquakes ≥5.0 magnitude. Most location uncertainties <35 km. Eleven more recent major or great earthquakes (magnitude ≥7.7) have been added for completeness through 2004; data from USGS National Earthquake Information Center at http://neic.usgs.gov/ , accessed January 4, 2005. An epicenter is the surface location of the first rupture on an earthquake fault. Symbols shown represent epicenters. For earthquakes larger than about magnitude 7.0, the size of the rupture zone, which can extend hundreds of kilometers from the epicenter, is larger than the symbols used on this map

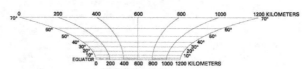

Depth to earthquake, in km	Magnitude of earthquake			
	5.0–5.9	6.0–6.9	7.0–7.9	≥8.0
<60	•	●	○	○
60–300	·	·	○	○
>300	·	○	○	○
Global average occurrence[1]	1319/yr	134/yr	17/yr	1/yr

[1]Earthquakes of magnitude <5 (not shown on map) are much more frequent, with ~13,000/yr in the 4.0–4.9 range alone. Data from USGS National Earthquake Information Center.

○ Earthquakes that occurred from 1750 to 1963 within stable plate interiors on continents—Data from A.C. Johnston (Center for Earthquake Research and Information, University of Memphis, written commun., 2002). Even though these epicenters do not meet the precise location criteria of Engdahl and Villaseñor (2002), they are plotted here to remind readers of the potentially hazardous earthquakes that are distant from known plate boundaries. Size of symbol proportional to earthquake magnitude

○ Notable pre-1900 earthquakes—Nos. 1, 2, 3, 6, and 7 (see table 3, on back)

0 200 400 600 800 1000 1200 KILOMETERS

MERCATOR PROJECTION
Scale 1:30 000 000 at the Equator
One centimeter equals 300 kilometers (~186 miles) at the Equator
One inch equals 473 miles (~762 kilometers) at the Equator

Suggested citation: Simkin, Tom, Tilling, R.I., Vogt, P.R., Kirby, S.H., Kimberly, Paul, and Stewart, D.B., 2006, This dynamic planet: World map of volcanoes, earthquakes, impact craters, and plate tectonics: U.S. Geological Survey Geologic Investigations Series Map I-2800, 1 two-sided sheet, scale 1:30,000,000.

Plate Tectonics

Divergent (sea-floor spreading) and transform fault boundaries—Red lines mark spreading centers where most of the world's volcanism takes place; thickness of lines indicates divergence rate, in four velocity ranges. White number is speed in millimeters per year (mm/yr) from DeMets and others (1994). The four spreading-rate ranges are <30 mm/yr; 30–59 mm/yr; 60–90 mm/yr; and >90 mm/yr. Thin black line marks the plate boundary, whether sea-floor spreading center or transform fault. On land, divergent boundaries are commonly diffuse zones (see interpretive map to the left); therefore, most are not shown. The only transform faults shown on land are those separating named plates

9 ◀— Plate motion—Data from Rice University Global Tectonics Group. Length of arrow is proportional to plate velocity, shown in millimeters per year. These approximate rates and directions are calculated from angular velocities with respect to hotspots, assumed to be relatively fixed in the mantle (see plate motion calculator at http://tectonics.rice.edu/hs3.html)

◁ **46** Plate convergence—More accurately known than "absolute" plate motion (above), convergence data are shown by arrows of uniform length showing direction and speed, in millimeters per year, relative to the plate across the boundary. Data from Charles DeMets (University of Wisconsin at Madison, written commun., 2003) and Bird (2003)

Legend for Figure 7.15

Part D: Summing Up

30. Complete the following summary sheet by placing an X in the appropriate box for each geologic process.

Geologic Process	Type of plate boundary		
	Divergent	Convergent	Transform fault
Creation of oceanic crust			
Destruction of oceanic crust			
Volcanism			
Creation of mountain ranges			
Compression			
Tension			
Shear			
Normal faulting			
Strike-slip faulting			
Reverse faulting			
Shallow EQs only			
Shallow and deep EQs			

31. Based on all the information above, what specific geographic location can be considered one of the most dangerous places on Earth? (*Hint*: See Figure 7.15. on previous pages)

LABORATORY EXERCISE VIII

EARTHQUAKES AND TSUNAMIS

INTRODUCTION

An earthquake is an episode of ground shaking caused by elastic waves propagating in the Earth, which were generated by a sudden release of slowly accumulated stress in rigid bedrock. This sudden release of stored energy results from an abrupt slippage of rock masses along a fault. The place where the slippage occurs is known as the hypocenter or focus of the earthquake, and the point on the surface vertically above the focus is the epicenter. See Figure 8.1.

OBJECTIVE

The objective of this lab is fourfold: 1) explore the physics of earthquake waves and their relationship to magnitude, intensity, and seismic risk analysis; 2) apply those concepts to an actual case study of the 1989 Loma Prieta Earthquake; 3) investigate a method of earthquake hazard "forecasting"; and 4) examine secondary effects of large earthquakes using the 2004 Indonesian Tsunami as a case study.

Part A: Earthquakes and Waves

The energy released at the hypocenter of an earthquake travels as several types of vibrations or waves that are transmitted through the Earth in all directions. Some waves travel through internal Earth and are known as "body waves." Others travel along the Earth's surface and are known as "surface waves." See Figure 8.1.

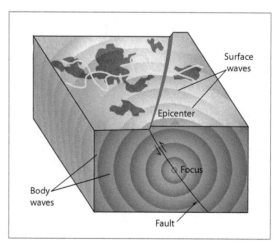

Figure 8.1 – Earthquakes and Waves

During any earthquake, two types of body waves are produced. The first is known as the "P-wave" or "primary wave," and the second is known as the "S-wave" or "secondary wave." P-waves propel themselves forward by compression and extension; this motion can be demonstrated by thumb-tacking a slinky to the wall and repeatedly pulling the slinky backward and forward. As such, P-waves are said to travel in a horizontal motion. S-waves, unlike P-waves, have the addition of vertical motion. Their method of travel can best be demonstrated by thumb-tacking a jump rope to the wall, then taking hold of the loose end and moving your hand up and down. See Figure 8.2. Note: Both the P-wave and the S-wave are generated and released from the hypocenter at exactly the same time, but reach their destination at different times due to the S-wave traveling at approximately half the speed of the P-wave.

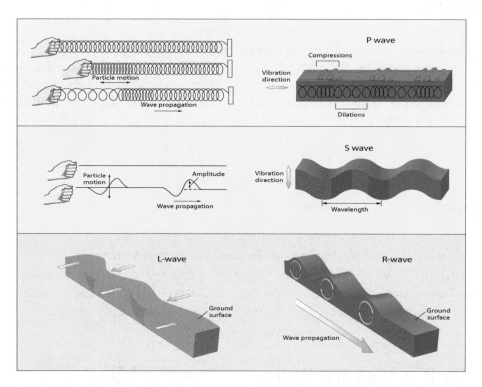

Figure 8.2 – Motion of Earthquake Waves

Another class of seismic waves is surface waves. These waves are generated as a result of the body waves smashing into Earth's surface. Surfaces waves consist of "Love waves" (travel forward but shake sideways) and "Rayleigh waves" (travel forward via a rolling motion). So the combined effect of both surface waves creates a very aggressive motion in both a horizontal and vertical direction. See Figure 8.2.

Figure 8.3 is a typical seismogram—a record of the ground's motion (Y-axis) as a function of time (X-axis)—marking the arrival of the P-waves, S-waves, and surface waves (L-waves and R-waves) into a measuring station called a seismograph. Use the seismogram in Figure 8.3 to answer the following questions.

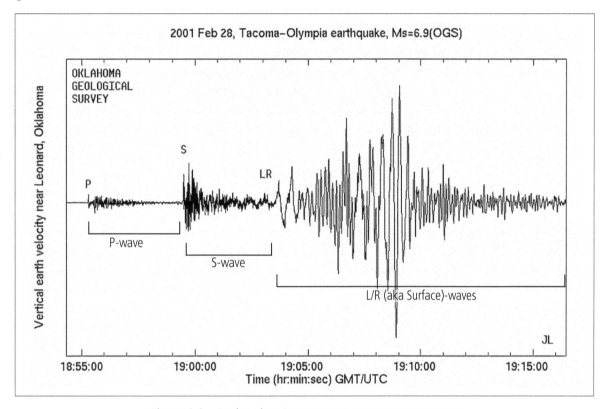

Figure 8.3 – Earthquake seismogram. SOURCE: THE IRIS CONSORTIUM.

1. Based on the seismogram above, which of these waves is the fastest? **P-wave**, **S-wave**, or **Surface wave** (L/R-wave) _____

2. Based on the seismogram above, which of these waves is the slowest? **P-wave**, **S-wave**, or **Surface wave** (L/R-wave) _____

3. Which of the three waves would you expect to cause the most damage? **P-wave**, **S-wave**, or **Surface wave** (L/R-wave) _____

    What's your evidence? _____

Part B: Epicenter, Magnitude, Intensity, and Seismic Risk

Epicenter

After an earthquake, seismologists are faced with the task of finding when and where the shaking began—meaning, they must locate the epicenter. Since most earthquake-prone areas are currently outfitted with seismographs to record the shaking during an earthquake, the location of an epicenter is usually determined in a matter of seconds, as the seismographs transmit their data to seismic laboratories in real time. If, however, you are interested in learning the science behind epicenter determination, investigate the concept of triangulation, then test your knowledge using the supplemental learning Section F at the end of this lab.

Magnitude

Magnitude is the term used to describe the highest amplitude of an earthquake wave produced during an earthquake. The Richter Magnitude Scale is a commonly used standardized system of amplitude measurement, and allows for comparison of different earthquakes around the world. However, due to inherent limitations associated with the Richter scale, scientists now use a more quantitative and accurate scale called Moment Magnitude, which involves examining the amplitude, the amount of slip, and the size of an area over which the slippage occurred. Both the Richter and Moment magnitude scales are logarithmic. This means that for every whole number increase in magnitude, there is a 10-fold increase in amplitude and a 32-fold increase in the amount of energy released by the earthquake (see table 8.A). It is important to note that the magnitude of an earthquake is the same for all areas affected by the earthquake. In essence, an earthquake will have *only* one magnitude.

Table 8.A – Earthquake magnitude, ground displacement, and energy release—they're all connected

Δ Magnitude	Δ Ground Displacement	Δ Energy
1	10x	~32x
0.5	3.2x	~5.5x
0.3	2x	~3x
0.1	1.3x	1.4x

Note: Δ = "change"

4. If an earthquake in Australia had a magnitude 5.0 and an earthquake in Indonesia had a magnitude 7.0, how much bigger was the amplitude of the second quake than the first? _____ x larger; How much more energy was released in the second quake than the first? _____ x more energy

5. Use Figure 8.4 to answer the following questions:

a. Approximately how many earthquakes measuring magnitude 2 are recorded each year?

_____

b. How many earthquakes measuring magnitude 7 or greater are measured each year?

_____

c. What is the relationship between increasing earthquake magnitude and the number of earthquakes that occur each year? _____

_____

d. Why do we not hear about most of the earthquakes that occur each year?

_____

_____

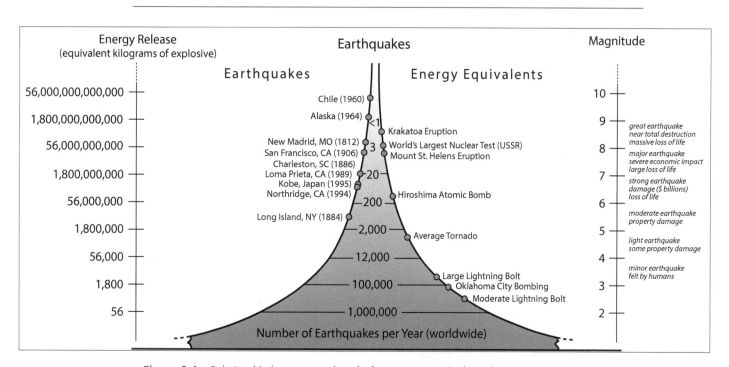

Figure 8.4 – Relationship between earthquake frequency, magnitude, and energy. THE IRIS CONSORTIUM

Intensity

In addition to measuring the size of an earthquake by means of seismographs, we also survey people who felt the earthquake and interpret their recollection of the level of intensity using the Modified Mercalli Intensity Scale (see Table 8.B). The intensity of an earthquake at a site is based on the observations of individuals during and after the earthquake. It represents the severity of the shaking, as perceived by those who experienced it, as well as observations of damage to structures, movement of stationary objects, and change in the Earth's surface as a result of geologic processes during the earthquake. As such, unlike magnitude where there can only be one per earthquake, one earthquake can have many different intensities.

Modified Mercalli Scale		Magnitude Scale
I	Detected only by sensitive instruments	1.5
II	Felt by few persons at rest, especially on upper floors; delicately suspended objects may swing	2
III	Felt noticeably indoors, but not always recognized as earthquake; standing autos rock slightly, vibration like passing truck	2.5
IV	Felt indoors by many, outdoors by few, at night some may awaken; dishes, windows, doors disturbed; autos rock noticeably	3
V	Felt by most people; some breakage of dishes, windows, and plaster; disturbance of tall objects	3.5 / 4
VI	Felt by all, many frightened and run outdoors; falling plaster and chimneys, damage small	4.5
VII	Everybody runs outdoors; damage to buildings varies depending on quality of construction; noticed by drivers of autos	5
VIII	Panel walls thrown out of frames; fall of walls, monuments, chimneys; sand and mud ejected; drivers of autos disturbed	5.5 / 6
IX	Buildings shifted off foundations, cracked, thrown out of plumb; ground cracked; underground pipes broken	6.5
X	Most masonry and frame structures destroyed; ground cracked, rails bent, landslides	7
XI	Few structures remain standing; bridges destroyed, fissures in ground, pipes broken, landslides, rails bent	7.5
XII	Damage total; waves seen on ground surface, lines of sight and level distorted, objects thrown up in air	8

Table 8.B – Modified Mercalli Scale.

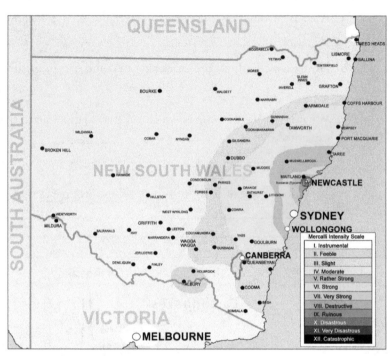

Figure 8.5 – 1989 Newcastle, Australia Intensity Map. ILLUSTRATION: BIDGEE

Isoseismal (intensity) maps, like Figure 8.5, show the distribution of seismic intensities associated with an earthquake. The greatest impact of earthquake waves is usually in the epicentral region, with progressively lower intensities occurring in nearly concentric zones outward from the epicenter. It must, however, be kept in mind that there are exceptions to this general rule, as the quality of construction and variation of geologic conditions affect the distribution of intensity.

6.   Given that a high percentage of intensity determination is based on interviews with survivors, as well as observations of damage to structures, what three non-geologic factors could significantly impair the geographic accuracy of the contour lines?

- _____

- _____

- _____

7. Use Figure 8.6 to create an isoseismal map for the December, 1872 Pacific Northwest Earthquake by drawing boundaries between the intensities. Then use the isoseismal map you created to answer the questions that follow.

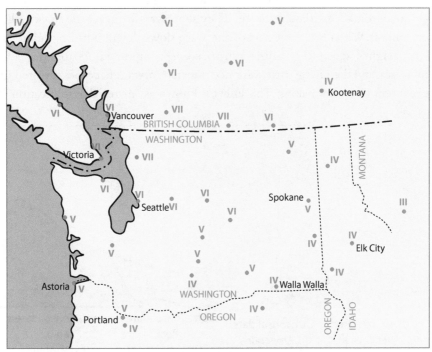

Figure 8.6 – Intensities for the December, 1872 Pacific Northwest Earthquake.

a. What was the maximum intensity for the 1872 Pacific Northwest Earthquake? _____

b. Shade in the area that represents the epicenter of the 1872 Pacific Northwest Earthquake.

c. While the intensity map in Figure 8.6 depicts no irregularities, oftentimes the expected "bulls-eye" pattern is disrupted with intensities that are "out of place." An example of this can be seen in Figure 8.7, which is an isoseismal map of the 1971 San Fernando Earthquake. Notice there are pockets of intensity VIII that are nearly on the boundary between the areas delineated as intensity VI and intensity VII. What are the two most likely causes for this localized increase in intensity? (Hint: One issue is geologic and the other is engineering.)

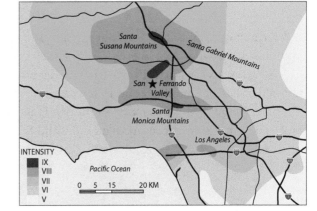

Figure 8.7 – Intensity map of the 1971 San Fernando Earthquake. USGS

*   _____

*   _____

Material Amplification

Earth materials of various types behave differently during earthquakes and the difference is related to their degree of consolidation and saturation. Seismic waves move forward faster in consolidated bedrock than through unconsolidated sediment or soil. They slow even further if the unconsolidated material has a high water content. When the front part of the wave slows down, but the middle and back continue forward at the original speed, the wavelengths become compressed. As the wavelengths shorten, the amplitude increases, and the energy that was once directed forward becomes directed upward, effectively increasing the intensity of the shaking. This effect is known as "material amplification." See Figure 8.8.

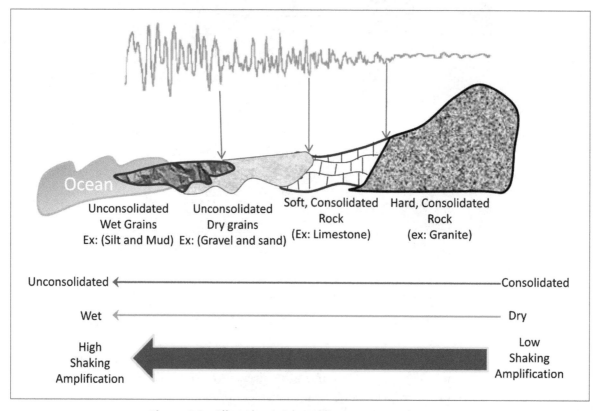

Figure 8.8 – Effect of material amplification on ground motion.

In determining seismic hazards within specific areas, geologists consider the potential for amplification of ground shaking by local surface and shallow subsurface geology, and the effects of that shaking on man-made structures. Of particular importance is the potential for liquefaction, where the ground "liquefies" due to the shaking, and slides out from under the structures it supports, leaving the city or town a heap of rubble. Figure 8.9 is a shake map of the San Francisco region and Figure 8.10 depicts the potential for liquefaction in the area. Examine figures 8.9 and 8.10, then answer the questions that follow.

8. On Figure 8.10, areas depicted by white are primarily underlain by bedrock, areas depicted by yellow and orange are primarily underlain by alluvium (sand, gravel, and cobbles), and areas depicted by pink and red are primarily underlain by mud.

    a. Comparing figures 8.9 and 8.10, what type of shallow, subsurface geology is most susceptible to liquefaction? _____

    b. What type is least susceptible to liquefaction? _____

    c. Explain the reason for this difference: _____

        _____

        _____

9.

    a. Comparing figures 8.9 and 8.10, which type of geology typically experiences the greatest shaking amplification? _____

    b. Which type of geology typically experiences the least shaking amplification?

        _____

10.

    a. Compare areas designated by red and yellow in Figure 8.10. What do you notice about the degree of development in those two areas with respect to each other? _____

        _____

    b. Why do you think there is less development in the red areas than the yellow areas?

        _____

        _____

        _____

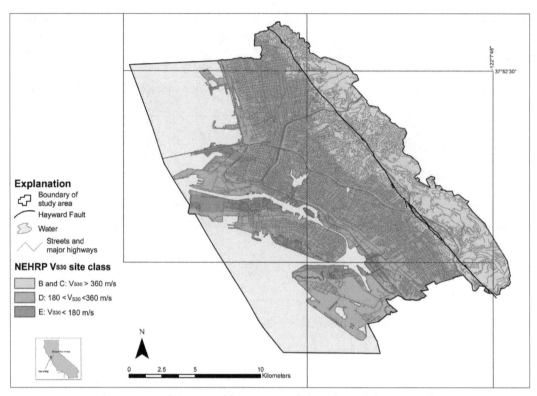

Figure 8.9 – Shaking amplification map of Alameda, Berkeley, Emeryville, Oakland, and Piedmont, California. USGS

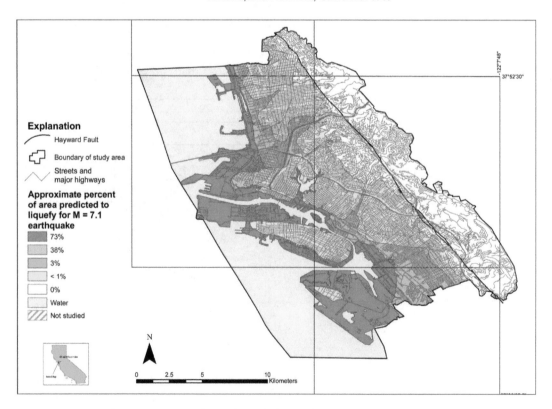

Figure 8.10 – Liquifaction hazard map of Alameda, Berkeley, Emeryville, Oakland, and Piedmont, California. USGS

Seismic Risk Maps

Seismic risk maps are based on the distribution and intensities of past earthquakes or on the probability of future earthquake occurrences (of a given ground motion in a given time period). National maps of earthquake shaking hazards provide information that helps to save lives and property by providing data for building codes, insurance rates, and hazard analysis. Buildings designed to withstand severe shaking are less likely to injure occupants. The hazard maps are also used by insurance companies, FEMA, EPA, and land-use planners.

Buildings are naturally designed to withstand high vertical forces, as they must contend with the force of gravity all the time. As such, it is the horizontal forces during an earthquake that usually cause catastrophic structural failure. The "shake maps," including the map below, depict hazard zones based on levels of horizontal shaking. Seismic hazard maps are probabilistic maps that are typically based on the anticipated degree of shaking expressed as ground acceleration, which is quantified as a percentage of the force of gravity (g or pga). The higher the number, the stronger the shaking. Use information in Figure 8.11 to answer the following questions.

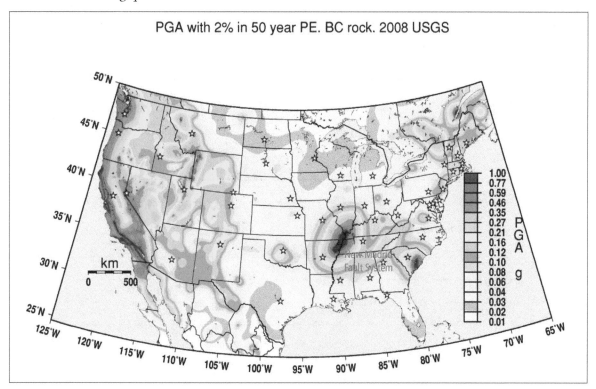

Figure 8.11 – Shake map for the contiguous 48 states.

11. Which areas of the country have the lowest hazard from earthquake shaking?

_____

12. Which areas of the country have the highest hazard from earthquake shaking?

_____

13. The geologic material on which a building rests plays a role in the amount of shaking that occurs during an earthquake. Which of the following foundation materials would most likely result in less shaking and safer building? **Artificial fill**, **poorly consolidated sediments**, **marine clays**, or **unweathered bedrock**. (Circle one)

14. The New Madrid Seismic Zone near Memphis, TN and St. Louis, MO experienced several large earthquakes during 1811 and 1812. However, unlike California, which has many smaller earthquakes in addition to infrequent large ones, earthquakes in this region happen on average approximately every 200 years, and are extremely destructive, as there has been no energy release for long periods of time. Due to the rarity of earthquake activity along the New Madrid Fault, there are relatively few people who are aware of or understand the severity of seismic risk in this region. This being the case, provide some reasons why an earthquake in this region could be more catastrophic than one in Los Angeles, with respect to the following:

   a. Building environments _____
   _____
   _____
   _____
   _____

   b. Risk perception of residents _____
   _____
   _____
   _____
   _____

   c. Frequency and magnitude of the quakes _____
   _____
   _____
   _____
   _____

Part C: Loma Prieta Case Study

On October 17, 1989, the San Francisco Bay region was shaken up by the Loma Prieta Earthquake, named after the small community located near the epicenter. San Francisco and Oakland were the focus of media attention because of their size and the amount of damage they sustained, but it's important to remember that they were 100 km away from the epicenter. Other smaller but closer communities, such as Watsonville and Santa Cruz, also suffered great damage and fatalities. This exercise looks at the geology of the earthquake and the geologic reasons for building failure in the San Francisco and Oakland region.

Geology of the San Andreas Fault System

Use Figure 8.12 to answer the following questions:

15. What is the approximate geographic orientation (compass direction) of the faults shown in Figure 8.12? _____

16. What large-scale geological, tectonic process is occurring in this location that accounts for the fault orientation and motion? (*Hint:* What type of fault is this and in what direction is each side of the fault moving with respect to the other side?) See Figure 8.13 for a visual aid.

_____

_____

_____

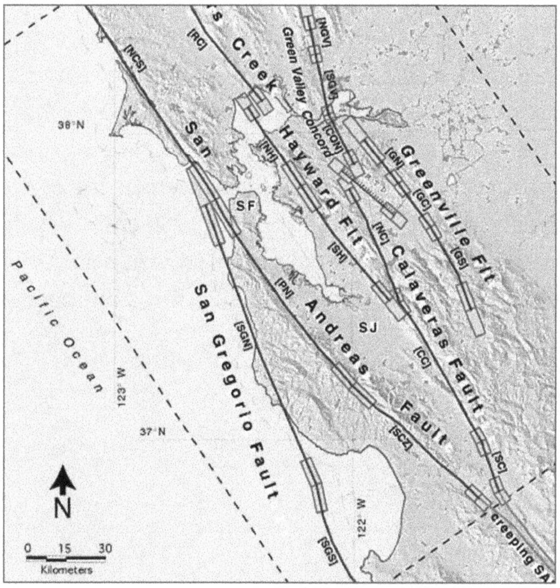

Figure 8.12 – Major branches of the San Andreas Fault System.

Figure 8.13 – Fault movement on the San Andreas.

Attenuation and Acceleration

Attenuation is the decrease in the amplitude of seismic waves as distance from the source increases. As a general rule, waves attenuate as they move away from the epicenter, but selected geologic conditions tend to cause local variations in amplitude, such that sites distant from the epicenter will show an amplification of ground motion.

Seismologists use sophisticated instruments to measure many different components of earthquake waves in an attempt to determine their affect on the local environment. An understanding of wave behavior with respect to its impact on the local geologic conditions is critical to establishing safe building codes, as the buildings must be able to sustain the forces exerted on the building from both the waves and the shaking sediment beneath them. During an earthquake, a small particle on Earth's surface will be moved back and forth haphazardly. This movement can be described by its changing position/unit time (displacement), changing velocity/unit time (velocity), or changing acceleration/unit time (acceleration).

Acceleration is the rate of change in velocity, i.e., how much the velocity changes in a unit time. You experience acceleration every time you increase or decrease the speed of your car by applying the gas pedal or brake respectively. When acceleration acts on a physical body, the body experiences the acceleration as a force. In the example with your car, your body experiences the force from the increasing velocity (depressing the gas pedal) by pushing you back into your seat, while it responds to the force of decreasing velocity (depressing the brake) by propelling you toward the windshield. Now picture the earthquake wave as the gas pedal/brake and the building as you in the car. The building will be pushed in multiple directions just as you would be pushed back and forth if the car were abruptly changing speeds. The faster the change in velocity, the stronger the force (e.g., steady braking vs. slamming on the brakes). If the acceleration (change in velocity per unit time) was not constant, but fluctuated throughout the period of acceleration, the largest value of the acceleration would be the "peak" acceleration.

While displacement, velocity, and acceleration are all ways to describe the changing position of material over time, acceleration is the most useful for our purposes because the building codes prescribe how much horizontal force buildings should be able to withstand during an earthquake. Accelerographs, which measure acceleration of the ground, record data that are typically presented as a percentage of the force of gravity (g). Buildings are designed to withstand the vertical force of gravity (1 g, 980 cm/sec/sec), hence the reason they are able to stand up under gravitational forces. They are not, however, necessarily designed to be moved sideways, and earthquakes generate some strong lateral forces against a building.

Figure 8.14 is a map showing horizontal acceleration from the Loma Prieta Earthquake at a number of sites in western California. Use this figure to answer the following questions:

17. A contour line has been drawn around the sites with a horizontal acceleration of 20% g.

a. What is the general compass orientation of the contour? _____

b. Any other contour line would replicate this orientation. How does its orientation compare with the orientation of the faults that are shown in Figure 8.12? _____

c. Had you drawn this line yourself (which I did and it took me a while because all I had were numbers instead of the color coded dots I supplied you with), you would see there is a large number of irregularities that make drawing this contour line a challenge. What main geological factor contributes to difficulties in drawing the irregular 20% g isoseismal line?

_____

d. What main engineering factor contributes to difficulties in drawing the irregular 20% g isoseismal line? _____

e. What happens to ground acceleration as you move progressively away from the epicenter?

_____

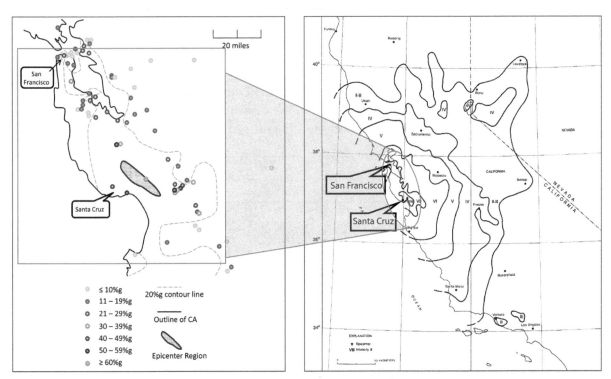

Figure 8.14 – Horizontal acceleration from the Loma Prieta Earthquake presented as % of gravitational force.

Figure 8.15 – Intensity map of 1989 Loma Prieta Earthquake region. USGS

f. What does this suggest about the role of distance from the epicenter in attenuating earthquake waves? _____

18. Figure 8.15 is an isoseismal map showing areas that sustained similar structural damage as measured by the Modified Mercalli Intensity Scale. Compare this figure to Figure 8.14 and describe the relationship between acceleration and structural damage.

_____

_____

_____

Earthquake Forecasting

Use Figure 8.16 to answer the following questions.

19. One earthquake prediction theory suggests that new events will fill in gaps where earthquakes have not been recently recorded. These gaps are presumed to be zones where stress is accumulating and will be released in a future earthquake.

a. Note the location of earthquakes and aftershocks prior to the Loma Prieta Earthquake in the top row of Figure 8.16. Compared to other locations along the fault, did the shallower depths beneath the Loma Prieta region appear to be in a period of stress build-up or relaxation from recent energy release? Explain your reasoning.

_____

_____

b. Now observe the location of the earthquake and aftershocks associated with the Loma Prieta Earthquake (bottom row of Figure 8.16). Does the Loma Prieta region appear to be in a period of stress build-up or relaxation from recent energy release? _____
Does this event support or contradict the filling-in-gaps theory? _____
Explain: _____

_____

_____

c. If this theory is indeed correct, using a red pen, draw 3 arrows on Figure 8.16 pointing to the 3 areas where the next "big one" on this fault is most likely to occur.

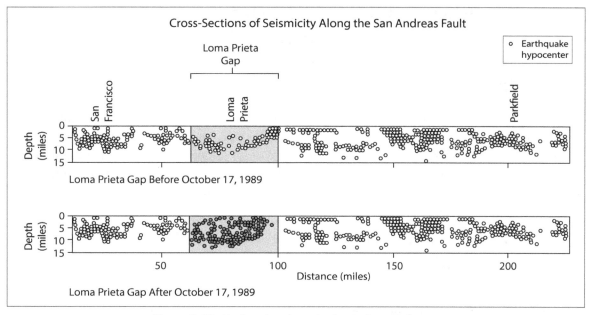

Figure 8.16 –Earthquakes along the San Andreas Fault. USGS

20. Figure 8.17 depicts all earthquakes magnitude ≥ 5.5 on the San Andreas Fault system from 1850 to 2000. This data was extrapolated to calculate the probability of a major earthquake on the San Andreas Fault system between the years 2003 and 2032. The likelihood of an earthquake ≥ 5.5 magnitude is represented by the colored region beneath the data tabs. Notice the places in which the probability is low and the places in which it increases. Is there a detectable pattern here? **Yes** or **No**? If yes, does it support or contradict the "seismic gap" theory?

Explain: _____

_____

_____

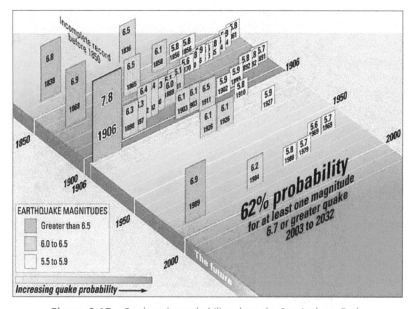

Figure 8.17 – Earthquake probability along the San Andreas Fault.

Part D: The Boxing Day Tsunami, 26 December 2004

The 2004 Indian Ocean earthquake was an undersea megathrust earthquake that occurred on December 26, 2004, with an epicenter off the west coast of Sumatra, Indonesia. The earthquake was caused by subduction and triggered a series of devastating tsunamis along the coasts of most landmasses bordering the Indian Ocean, killing nearly 250,000 people in eleven countries, and inundating coastal communities with waves up to 30 meters (100 feet) high. It was one of the deadliest natural disasters in recorded history. Indonesia, Sri Lanka, India, and Thailand were the hardest hit. With a magnitude of between 9.1 and 9.3, it is the second-largest earthquake ever recorded on a seismograph. In addition, this earthquake had the longest duration of movement ever observed, between 8.3 and 10 minutes. It caused the entire planet to vibrate as much as 1 cm and triggered other earthquakes as far away as Alaska.

21. As mentioned above, the type of plate boundary where this earthquake occurred is a subduction zone. What is the relative movement of the two plates (divergent, convergent, strike-slip)?

   _____

22. Give one major example on the globe where this type of boundary exists. (*Hint:* Think about where most land-based volcanic activity on earth is located).

   _____

When the actual slippage occurred as a result of the megathrust earthquake, the uplifted ocean floor subsequently displaced a large volume of ocean water upward, which then raced across the ocean at about 600 kilometers per hour. Use Figure 8.18, which depicts tsunami travel time contours (in hours) spreading out from the epicenter (red star), to answer the questions that follow.

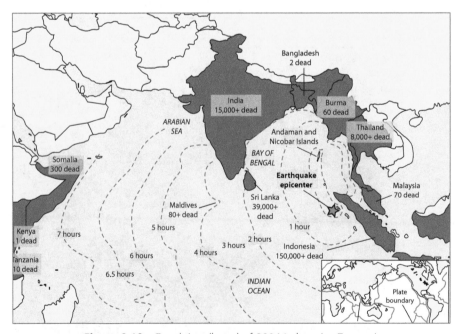

Figure 8.18 – Travel time (hours) of 2004 Indonesian Tsunami.

23. How much time did Indonesians have to react to the news that a tsunami was on its way?

_____

Is this enough time to evacuate highly populated, low-lying coastal areas? **Yes** or **No**?

24. If the people in the countries closest to the epicenter had been educated with respect to earthquake hazards and what to do in the event of a powerful earthquake (i.e., immediately move to high ground in case a tsunami develops), would that have changed the number of casualties? **Yes** or **No**? Explain: _____

_____

_____

25. How much time would Indians have been given if a tsunami warning was issued immediately after the earthquake? _____

26. Name several reasons why the number of casualties was so high, with deaths recorded as far away as Africa.

- _____

_____

- _____

_____

- _____

_____

27. Based on your answers to #26, were most of these deaths preventable? **Yes** or **No**?

28. What are the two safest places to go in the event a tsunami occurs?

- _____

- _____

Part E: Magnitude 9.0 Earthquake and Subsequent Tsunami on East Coast of Japan, 11 March 2011

The magnitude (M) 9.0 Tohoku earthquake on March 11, 2011, which occurred near the northeast coast of Honshu, Japan, resulted from thrust faulting on or near the subduction zone plate boundary between the Pacific and North American plates. At the latitude of this earthquake, the Pacific plate moves approximately westwards with respect to the North American plate at a rate of 83 mm/yr, and begins its westward descent beneath Japan at the Japan Trench. See Figure 8.19.

The Japan Trench subduction zone has hosted nine events of magnitude 7 or greater since 1973. Large offshore earthquakes have occurred in the same subduction zone in 1611, 1896 and 1933 that each produced devastating tsunami waves on the Sanriku coast of Pacific NE Japan. That coastline is particularly vulnerable to tsunami waves because it

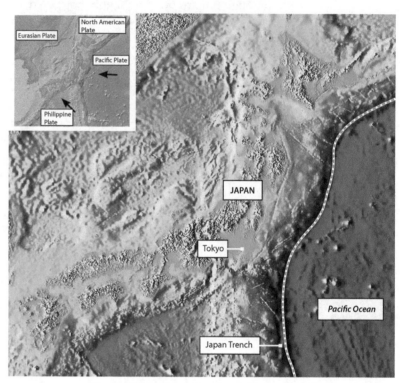

Figure 8.19 – Japan Trench subduction zone

has many deep coastal embayments that amplify tsunami waves and cause great wave inundations. The M 7.6 subduction earthquake of 1896 created tsunami waves as high 38 m and a reported death toll of 27,000. The M 8.6 earthquake of March 2, 1933 produced tsunami waves as high as 29 m on the Sanriku coast and caused more than 3000 fatalities.

The March 11, 2011 earthquake far surpassed other post-1900 plate-boundary thrust-fault earthquakes in the southern Japan Trench, none of which attained M 8. A predecessor may have occurred on July 13, 869, when the Sendai area was swept by a large tsunami that Japanese scientists have identified from written records and a sand sheet. (Courtesy of the U.S. Geological Survey website.)

Use Figures 8.19 and 8.20 to answer the following questions.

29. What type of plate boundary marks this location? **Divergent**, **Convergent**, or **Strike-slip**?

_____

30. What type of plates are interacting at this plate boundary? **Oceanic/Oceanic**; **Oceanic/Continental** or **Continental/Continental**?

_____

31. Figure 8.20 depicts the historic seismicity along the east coast of Japan between 1990 and the March 11, 2011 earthquake. Notice that the earthquake foci become progressively deeper in the westward direction from the trench. Why are the earthquake depths changing in this fashion?

_____

_____

_____

_____

32. *To be done at home and e-mailed in.* Among the most devastating consequences of the March 11, 2011 earthquake and Tsunami was the destruction of the Fukushima Nuclear Power Facility, which created what is believed to be the worst nuclear accident since the Chernobyl disaster in 1986. Take a few minutes to research this incident on the internet. After learning about the events that led up to this disaster, use this information in conjunction with what know about the historical behavior of earthquakes and Tsunamis in this region to answer the following questions:

a. List at least 3 reasons why the power plant was an accident waiting to happen.

b. Was this catastrophe preventable? Explain.

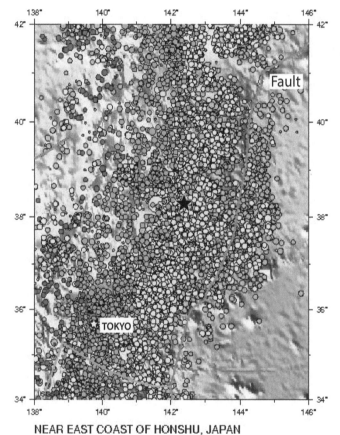

NEAR EAST COAST OF HONSHU, JAPAN

2011 03 11 05:46:24 UTC 38.29N 142.37E Depth: 30.0 km

Seismicity 1990 to Present

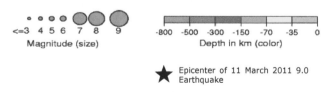

Magnitude (size)   Depth in km (color)

Epicenter of 11 March 2011 9.0 Earthquake

Figure 8.20 – Historic Seismicity on East Coast of Japan: 1990–March 11, 2011

Part F: Supplemental Learning Section

Epicenter

After an earthquake, seismologists are faced with the task of finding when and where the shaking began, meaning they must locate the epicenter. Keeping track of epicenter locations in earthquake prone regions is essential to understanding the fault behavior, and therefore is a central part of earthquake forecasting, land-use assessment, and building code requirements. Since most earthquake-prone areas are currently outfitted with seismographs to record the shaking during an earthquake, the location of an epicenter is usually determined in a matter of seconds, as the seismographs transmit their data to seismic laboratories in real time. However, it is important to understand the physics behind this process, as epicenter identification, along with anything linked to earthquakes, is directly correlated to wave behavior.

It is now known that earthquake waves travel at a fixed speed, depending on the type of rock the wave is traveling through. The denser the material, the more confined the wave and the faster the wave can propel forward. Laboratory experiments have carefully documented the speed at which each of the different waves travel through the various rocks making up both the surface and internal Earth. Those experiments permitted the development of a universal graph that demonstrates the distance each of the waves can travel throughout Earth per unit time, based on the rock types comprising the planet (Figure 8.21). As such, if the travel time of a wave from its generation to the time it reaches a seismograph is known, the universal graph can be used to determine the distance from the epicenter to the seismograph. Note that since the creation of this universal graph, it has been used to accurately identify the epicenter locations of thousands of earthquakes around the world and is therefore supported by field data.

Exercise 1: Locating the Epicenter of an Earthquake

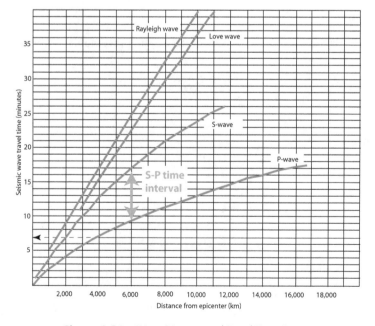

1. Examine Figure 8.21. Why are the lines of the P- and S-waves curved, whereas the line for the surface waves is straight? _____

_____

_____

_____

_____

_____

Figure 8.21 – Wave Distance and Travel Time Curves

As previously discussed, in order to use this graph, the travel time of the wave must be known. Yet when the P-wave arrives at a seismograph, the only information available is the arrival time of the wave—which does not tell us how long the wave has been traveling. In order to rectify this problem, scientists use what is known as the S-P time interval, which is the amount of time between the arrival of the P-wave and the arrival of the S-wave. Using this relationship, we have a known wave travel time interval and can therefore use the graph to calculate the distance of the seismograph from the epicenter. Figure 8.22 is a simplified version of the universal graph in which the S-P time interval has been depicted on one curve. For example, if it was determined that 30 seconds passed between the arrival of the P-wave and the arrival of the S-wave, on a traditional graph it would be necessary to find the spot between the P and S wave curves that represents 30 seconds (see blue line on Figure 8.22), then read down to the Y-axis to determine the distance from the epicenter (300 km). However, the addition of the S-P time interval curve allows you to start at 30 seconds on the Y-axis, read across till you reach the S-P time interval curve, then read down to the X-axis; again getting 300 km as the distance from the epicenter (see red lines on Figure 8.22). Using the simplified S-P time interval curve will allow you to quickly determine the distance from the epicenter for any seismogram.

2. What happens to the S-P time interval as the distance from the epicenter increases?

    _____

3. Refer back to Figure 8.3.

    a. What is the arrival time of the P-wave?
        _____

    b. What is the arrival time of the S-wave?
        _____

    c. What is the S-P time interval?
        _____ minutes

    d. Using Figure 8.21, determine how far the seismograph that produced the seismogram in Figure 8.3 was from the epicenter? _____km

4. While going through the process described above will tell you the distance of the seismograph from the epicenter, it does not tell you the actual location of the epicenter. Why not? _____
    _____
    _____
    _____

5. What additional information do you need to determine the actual location of the epicenter?
    _____
    _____

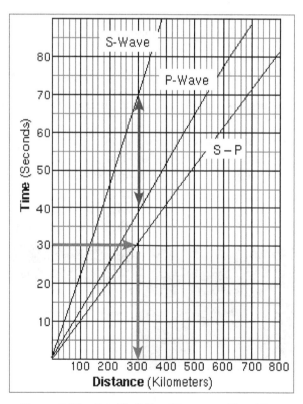

Figure 8.22 – Simplified Wave Distance and Travel Time Curves Directly Showing S-P Time Interval

Once you obtain the distance of three different seismographs surrounding the epicenter, you can triangulate to find the actual epicenter location. This can be done by following the steps below:

I.   Obtain a map of the region you are working with.

II. Plot the locations of the three seismographs on your map and label them accordingly.

III. Using the bar scale on the map, open a compass the width representative of the distance from seismograph 1 to the epicenter.

IV. Put the sharp end of your compass on the location of seismograph 1, and draw a full circle around that location. The radius of that circle is the distance of seismograph 1 from the epicenter.

Figure 8.23 – Example of Triangulation

V.   Repeat steps III and IV, using the distances from seismographs 2 and 3, respectively.

The place where all three circles intersect is the location of your epicenter! See Figure 8.23.

6.   Use the three seismograms on the following page, the corresponding map (Figure 8.24), and the simplified S-P time interval curve (Figure 8.22) to locate the epicenter of an earthquake. Note: Only the P and S waves are visible on the seismograms. Surface waves have been removed for clarity.

Step 1: Locate the distance of the 3 seismographs from the epicenter

Station	S–P Interval (seconds)	Epicentral Distance (km)
Chichuahua		
Mazatlan		
Rosarito		

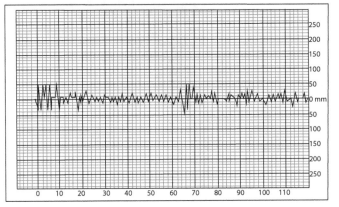

Chihuahua Seismic Station

Mazatlan Seismic Station

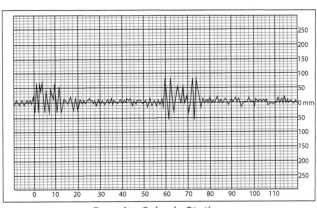

Rosarito Seismic Station

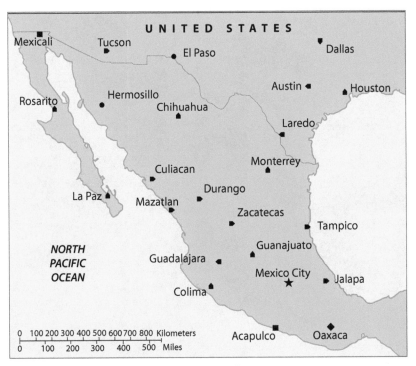

Figure 8.24 – Triangulation Map

Step 2: Triangulate

Where is the epicenter located?

_____

RIVERS AND FLOODING

INTRODUCTION

"Flooding" is a term used to describe a situation in which land that is not usually submerged becomes submerged. Flooding may result from a variety of causes including dam failures, high tides along coastal and estuarine areas (usually related to extreme weather events—like storms), melting of glacial ice during volcanic eruptions or the onset of Spring, and excessive precipitation and/or snow melt. Rivers flood when the volume of water flowing through the river exceeds its capacity and overflows the banks. It's important to keep in mind that flooding is a normal, natural and necessary part of river evolution, and something we should find ways to safely live with instead of trying to prevent the process.

Rivers are one of the most complicated systems on the surface and therefore have a wide range of variables that must be evaluated if we are to get an accurate understanding of the river's behavior and its expected pattern of flooding. Below is a brief introduction to the way hydrologists approach measuring and modeling river systems as well as the factors that contribute to flooding and flood damage.

Figure 9.1a – Organization of a Drainage Basin

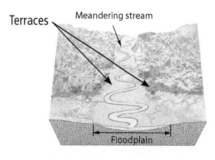

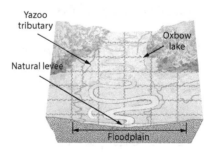

Figure 9.1b – River and Flood features

175

Rivers and Flood Features—See Figures 9.1a and 9.1b

A. **River/stream channel**: The topographic low area that contains the normal water flow.

B. **Drainage Network and Drainage Basin**: An interconnected system, consisting of the headwater channels, tributaries, and a trunk stream. The specific area the network drains is called a "drainage basin."

C. **Floodplain**: Land adjacent to the streambed that is commonly flooded and thus has been eroded to a relatively flat surface.

D. **Levee**: Topographic high area parallel to the river bank either man-made or formed naturally by deposition of sediment when the river floods.

E. **Meanders**: Bends in the river not related to structural control.

F. **Terraces**: Staircase-like topography adjacent to a river representing a series of historic floodplains and indicative of a rejuvenated stream.

G. **Drainage density**: A measure of the density (or number per area) of streams in an area; related to the local climate, topography, and age of a river system.

H. **Indications of flooding**: leaf litter in trees, ripple marks on floodplains, washed-over grass, high water lines on rock or buildings.

River Measurements

A. **River Stage**: Height of water in the river/stream with respect to an established reference datum (typically measured in feet).

B. **Cross-Sectional Area (A)**: The area of a transverse slice (slice perpendicular to stream flow) across the creek or river. Cross-sectional area can be calculated by multiplying the average depth of the stream by the width of the stream at the point of interest.

C. **Velocity (v)**: Speed of the water within the channel in the down-river direction (typically measured in feet per second or meters per second). Velocity may vary in different parts of the stream channel causing deposition in some areas and erosion in others. To calculate, divide distance by time ($v = d/t$).

D. **Flow or Discharge (Q)**: Volume of water passing a given point in the river per unit measure of time. Typically measured in cubic feet per second or cubic meters per second. To calculate, determine the cross-sectional area of the river bed. Multiply the cross-sectional area of the river bed by the velocity, making sure that you start with units that match (Q = Av).

Graphic Representations of Surface Water Measurements

A. HYDROGRAPHS

Hydrographs are a graphic representation of discharge versus time for a stream. They are created by measuring river discharge at a given geographic location over time. From hydrographs, we can determine:

Baseflow: the average flow in a river over time (during non-flood stage).

Response time: the duration of time between the start of a rainfall event and noticeable effects with respect to river stage (i.e., rise in river level).

Peak flow: the peak discharge of a river in response to rainfall.

Recovery time: how quickly the river returns to baseflow conditions following the cessation of rainfall.

B. RAINFALL INTENSITY GRAPHS

Rainfall intensity graphs show the volume of rainfall versus time (typically as inches per hour). These graphs are often overlaid onto hydrographs to evaluate the relationship between rainfall and river stage.

Rivers and Flooding

A. FLOOD FREQUENCY

Measured as the % chance EACH YEAR that a flood of a given magnitude will occur.

Example 1: 50-year flood—there is a 1 in 50 chance (or 2% chance) each year that a flood of a given magnitude will occur. This does *not* mean that a flood of this magnitude will only occur once every 50 years. The magnitude of the 50-year flood (measured as a river stage or discharge) is different for each river or drainage basin.

Example 2: 20-year flood—there is a 1 in 20 chance (or 5% chance) each year that a flood of a given magnitude will occur.

B. FACTORS THAT AFFECT THE FREQUENCY AND INTENSITY OF FLOODING:
   • Topography (steepness of land)
   • Size of floodplain
   • Size of watershed and location within watershed
   • Duration of rainfall

- Areal extent of rainfall
- Intensity of rainfall (for example - 4 inches in 1 hour or 1 inch over 4 hours)
- Infiltration capacity of the soil (how quickly the water can soak in)
- Wetness of the soil (damp soil soaks up water faster than completely dry soil), often related to climate (arid vs. humid)
- Time since the last precipitation event
- Tidal fluctuations in near coastal/estuarine environments (i.e., Potomac River at Georgetown)

C. FACTORS CONTRIBUTING TO, OR MITIGATING, FLOOD DAMAGE
- Land use in close proximity to rivers
- Urban planning (tracking water levels, evacuation and response plans)
- Use of flood control features (levees, flood gates, sand bags, etc.)
- Storm water retention ponds
- Storm sewers
- Amount of impermeable surface (asphalt, concrete, rooftops, etc.)
- Modification of natural flood protection

Rivers And Humans: - Why Do We Care?

- Land-use planning
- Design of storm sewers and storm water catchment basins
- Flood planning and predictions
- Ecological habitat (in-stream and wetland)
- Water quality (sediment)

OBJECTIVE

The objective of this lab is three-fold: 1) Introduce you the mechanics of different types of natural floods, 2) examine the anthropogenic contribution to flooding, and 3) investigate ways to mitigate the hazards caused by flooding events.

Part A: Flood Frequency and Major Floods

Where flood records are available, computations of flood frequency are based on peak annual floods (the maximum discharge for the year as measured at a specific station). Flood frequency is expressed as a recurrence interval, which is the likelihood that a flood of a given discharge can be expected to occur X times within a fixed time interval. The recurrence interval (T, in years) for a flood of a given discharge is determined by the formula below:

$$T = (n + 1)/m, \text{ where}$$
n equals the number of years of record and
m is the rank or order (1 = greatest flood and so on) of the annual peak flood discharge.

Exercise 1: Flood Frequency of the fictitious "Risky River"

Tables 9.A and 9.B contain data sets of Peak Flood discharge for "Risky River." Each data set spans 15 years of recording. Using one data set at a time, estimate the likely discharge for Risky River by following the steps below:

1. Rank the peak flood discharge for data set #1 in order of magnitude, starting with 1 for the largest and ending with 15 for the smallest. Write these results in the "rank" column.

2. Use the formula $T = (n + 1)/m$ and determine the recurrence interval for each of the 15 floods in data set #1. Write the results for each year in the "recurrence interval" column.

3. Repeat this procedure for data set #2.

Risky River Data Set 1

Year	Peak Flood Discharge (cfs)	Rank	Recurrence Interval
1949	165		
1950	235		
1951	198		
1952	183		
1953	170		
1954	152		
1955	139		
1956	220		
1957	169		
1958	162		
1959	257		
1960	178		
1961	189		
1962	200		
1963	173		

Table 9.A

Risky River Data Set 2

Year	Peak Flood Discharge (cfs)	Rank	Recurrence Interval
1985	499		
1986	387		
1987	635		
1988	585		
1989	458		
1990	312		
1991	817		
1992	525		
1993	502		
1994	463		
1995	208		
1996	315		
1997	560		
1998	430		
1999	708		

Table 9.B

Figure 9.2 depicts the peak flood discharge (given on the table) plotted against the recurrence interval you calculated for each of the data sets on the previous page. The graphs on the left show the data from the two time periods as a logarithmic curve, which is the way it plots on a normal grid, while the graphs on the right depict the two data sets on a logarithmic grid so that the curve is straightened out allowing you to view the data more easily. As such, the top two graphs depict the same data and the bottom two graphs depict the same data, just presented two different ways. A best-fit curve has been applied to the data, and the equation for each curve is displayed on the graph.

4. Based on the graphical representation data of set 1, what is the predicted discharge for a 100-yr flood? _____ cfs   (*Hint:* ln(100) = 4.60517)

5. Based on the graphical representation of data set 2, what is the predicted discharge for a 100-yr flood? _____ cfs   (*Hint:* ln(100) = 4.60517)

6. How much difference is there between the prediction from the first data set and the prediction from the second data set? _____ cfs

Data Sets Graphed on Normal Grid   **Data Sets Graphed on Logarithmic Grid**

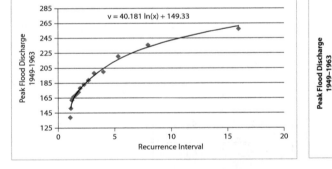
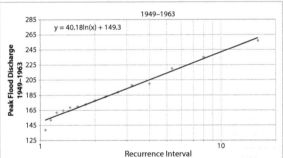
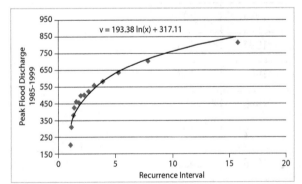
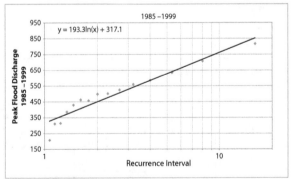

Figure 9.2

7. List some likely anthropogenic reasons for the increased discharge and therefore higher flood prediction in data set 2.

    - _____

    - _____

    - _____

8. Based on your answers above, would you say using data sets to extrapolate far into the future is an accurate and useful method for land use planners? **Yes** or **No**?

    Explain: _____

    _____

    _____

9. How might your answer to the previous question affect your decision to buy a home adjacent to but just outside the 100-year flood line?

    _____

    _____

    _____

Exercise 2: Large Floods in the US

Data in Table 9.C on the following page show damages suffered and deaths due to flooding in the US throughout the 20th century. Use Table 9.C to answer the following questions:

10. On the graph associated with Table 9.C, use dots to plot the monetary flood loss data for each decade. What is the general trend in flood loss in the US between 1900 and 2000?

    _____

11. On the same graph, use Xs to plot the US population at the end of each decade.

12. Assuming the data you graphed up from Table 9.C was taken from an area that used traditional physical barrier flood control measures such as levees and dams to protect the growing population, would you agree or disagree with the statement "attempting to prevent rivers from flooding through the use of physical barriers such as dams and levees have been effective in reducing flood damage." **Agree or Disagree?**

    Explain your answer: _____

    _____

    _____

    _____

Table 9.C – 20th century damages and deaths due to flooding

Decade	US Population (Millions)	Flood Damages (Billions of $)	Flood Fatalities
1900–1909	92	13.3	204
1910–1919	105.7	31.2	958
1920–1929	122.8	27.9	1048
1930–1939	131.7	41.2	929
1940–1949	150.7	32.6	619
1950–1959	179.3	48.2	791
1960–1969	203.2	30.5	754
1970–1979	226.5	68.2	1806
1980–1989	248.7	46.5	1123
1990–1999	281.4	77.4	994

*Population data from US Census Bureau; Flood Damages from noaa.gov; Flood Fatalities from weather.gov

*Flood damage figures adjusted to 2007 dollars

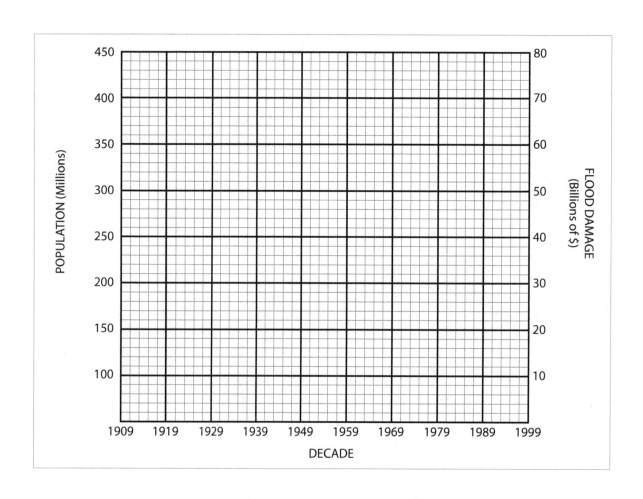

Exercise 3: Discharge and Area Ratio

The amount of damage that occurs when a river floods not only depends on how much water overflows the banks, but also on the size of the area the water covers. The same magnitude flood in two different locations can cause much more damage in a smaller area—as the water is more confined laterally resulting in higher flood levels. Alternatively, if that same amount of water spread over a larger area, the floodwater levels would be significantly lower. By the same token, if River X produces a "larger flood" than River Y, (i.e., flood with a higher discharge), the area surrounding River X may still suffer less damage than the area surrounding River Y, if the floodwater per unit area around River X is lower than the floodwater per unit area around River Y. The relationship between flood discharge and drainage basin size is referred to as Flood Intensity.

13. Use the data in Table 9.D on the following page to calculate the flood intensity for each category of rivers listed. Record your calculations in the right hand column in the space provided.

14. Plot the drainage basin area against the flood intensity, for all but the last 3 rivers, on the associated graph on the following page. The last 3 rivers follow the same trend, but because of their larger numbers, require a sizably larger graph than the one provided.

15. Based on your calculations and the resulting plot, what is the general relationship between flood intensity and drainage basin area—i.e., with increasing area of the drainage basin, is there an increase or decrease in the flood intensity? _____

Table 9.D

River	Drainage Basin Area (Mi²)	Maximum Discharge (ft³/S)	Flood Intensity Discharge/Area ((ft³/S)/Mi²)
Category A (10–35 Mi²)	22	87227	
Category B (100–200 Mi²)	168	201294	
Category C (400–600 Mi²)	575	347496	
Category D (700–1000 Mi²)	815	589755	
Category E (1500–5000 Mi²)	3112	752202	
Category F (6000–10000 Mi²)	9220	1306643	
Category G	389963	3884613	
Category H	938228	6709787	
Category I (Amazon)	1791514	12360133	

★The data for each category above represent an average of several rivers throughout the world. The range of drainage basin sizes measured to obtain the category average is listed beside the category label.

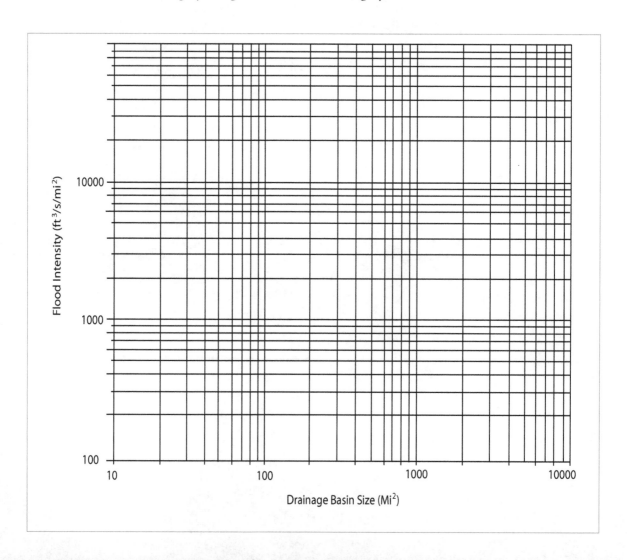

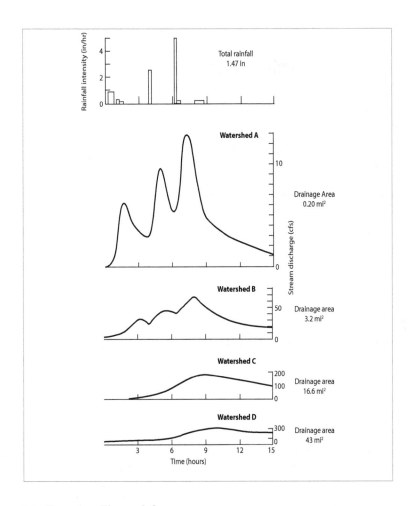

Figure 9.3 – Changes in Hydrograph shape in a series of Stations along the Sleepers River, near Danville, Vermont.
DATA FROM DUNN AND LEOPOLD, 1978.

16. Examine Figure 9.3.

a. For a given rainfall event, how does the stream discharge vary between larger and smaller watersheds? _____

_____

b. Calculate the flood intensity for each of the watersheds in Figure 9.3.
   watershed A. _____; watershed B. _____
   watershed C. _____; watershed D. _____

c. For a given rainfall event, how does the flood intensity vary between larger and smaller watersheds?

_____

_____

17. Based on this data, does a larger discharge necessarily mean a worse flood? **Yes** or **No**?

Explain your answer: _____

_____

_____

_____

Part B: Case Study
Cumberland River Flooding, Nashville, TN, 2010

Rivers throughout middle Tennessee crested at record high levels on May 2 and 3, 2010. They exceeded previous highs at many stream gages by as much as 14 feet, according to preliminary estimates released by the US Geological Survey (USGS). The highest flood levels were recorded on May 2 and 3, from Nashville west toward Jackson, extending about 40 miles north and south of Interstate 40, and affecting major tributaries to the Cumberland and Tennessee rivers.

The flood peak on the Cumberland River in downtown Nashville ranks as only the tenth highest in more than 200 years of records at that site. This peak was, however, the highest observed during the past 73 years in which much of the basin upstream of Nashville was regulated by several large flood-control reservoirs. During this flood, at least four major tributaries to the lower Cumberland River met or exceeded warning levels established by the National Weather Service (NWS) for major flooding conditions.

"Most tributaries to the lower Cumberland River had flows with only a 1 in 500 chance in any given year, causing the lower Cumberland to flood with a severity that was almost entirely unexpected," according to the USGS Tennessee Water Science Center. "That a regulated river like the Cumberland could have such high flooding is unusual and is a testament to the severity of this event. The extreme rainfall in tributaries that enter the Cumberland River downstream from the flood storage area made this a very difficult event to regulate."

Notable Floods on the Cumberland River, Nashville, TN

Year	Discharge	Gage Height (MSL)
1793	193,067 (estimated)	426.85
1808	174,531 (estimated)	422.17
1847	178,238 (estimated)	423.07
1882	179,886 (estimated)	423.47
1927	203,000	424.37
1937	186,000	422.07

A View of Some Notable Past Floods

The Great Ohio River Flood of January–February 1937 surpassed all prior floods during the previous 175 years of modern occupancy of the Ohio River Valley and geological evidence suggests the 1937 flood outdid any previous flood. NOAA.

Tennessee River Flood, Chattanooga, Tennessee, 1917. RG 82, Department of Conservation Photograph Collection. TENNESSEE STATE LIBRARY AND ARCHIVES.

The following data were acquired from a stream gaging station located on the Cumberland River in Davidson County, TN, at an elevation of 368.17 feet above mean sea level (msl). The peak flood stage (recorded number of feet above local baseflow height) of the 2010 flood at this gage was 51.86 ft, which means that the elevation of the flood at this point was 420.03 feet. Any structure in the immediate area of the gaging station and lower than this elevation would have suffered flood damage. For your reference, at this gaging station, the river is considered to be in "flood stage" at 40 ft or 408.17 ft above sea level.

In this exercise, we examine the extent, frequency, and causes of some river floods, and ways to reduce loss of life and property.

18. Using the data in Table 9.E on the following page, plot the recurrence interval against elevation on the graph provided and draw a best-fit line for frequency of floods at Nashville, TN.

19. What are the expected elevations of the following floods at the Nashville gaging station as determined from the graph you just plotted?

    20-year flood: _____

    75-year flood: _____

20. Figure 9.3 is a graph of floodwater elevation vs. discharge for the Cumberland River at Nashville, TN between the years 1815–2010. What trend do you see between stage level and discharge?

    _____

21. One main anthropogenic factor contributing to this trend is increased regulation by the Army Corps of Engineers. What's the other? _____

22. Discharge can be obtained for a flood of a given frequency by using information on the flood frequency-elevation plot (obtained from the data in Table 9.E), and the Stage-level–Discharge plot (Figure 9.4). What is the expected discharge at Nashville for:

    a. A 20-year flood? _____

    b. A 75-year flood? _____

    c. A 125-year flood? _____

Table 9.E – Recurrence Interval, Stage, and Discharge at Nashville, TN

Recurrence Interval at gaging station (yr)	Flood Stage	Elevation above mean Sea Level Stage (Ft)	Discharge (cfs)
2	38.6	406.77	110,000
5	44.8	412.97	137,500
10	48.4	416.57	154,500
30	52.7	420.87	175,100
50	55.6	423.77	189,900
100	58.4	426.57	204,300
150	61.1	429.27	218,500
200	64.6	432.77	237,000

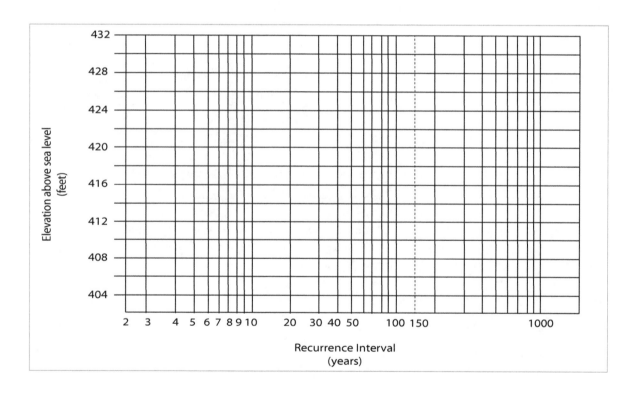

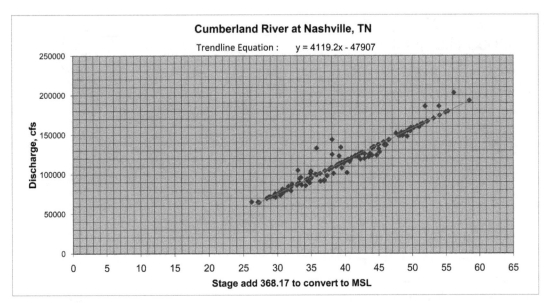

Figure 9.4

23. The May 2–3, 2010 event had a peak flood elevation of 51.86 ft and a peak discharge of 186,000 cfs. Using the data from this gaging station, calculate how often a flood of this magnitude can be expected based on:

a. The flood elevation _____

b. The discharge _____

c. Name a few reasons why the recurrence intervals you calculated in a and b above are different from each other, even though the data came from the same flood?

_____

_____

_____

d. Is flood data useful for "predicting" or "forecasting" (circle one)? Explain:

_____

_____

_____

24. If you visit a flood area after the water recedes, what indications might there be on buildings, trees, and land surfaces to indicate the level of the flood maximum?

• _____

• _____

• _____

Part C: Flash Flooding on the Fictitious Risky River

On the evening of April 12, 1978, flash floods in conjunction with erosion, deposition, and mass movement devastated the cities, towns, and the landscape surrounding Risky River. Following a few hours of heavy rainfall, there were 202 deaths and $50 million in damage mainly on Risky River between Lakeside to the east and Crushing Canyon to the west, approximately 8 miles west of River City (see Figure 9.5). The mountainous terrain in combination with the unusually high precipitation rates resulted in significant excess runoff and the Risky River quickly reached flood stage. Peak discharges exceeded the 100-year flood discharges at several locations.

The primary objective of this exercise is to understand the nature and causes of flash flooding in mountainous terrains. The secondary objective is to investigate the impacts of flooding and associated water erosion and sediment deposition.

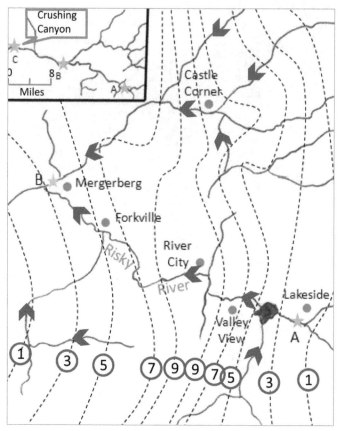

Figure 9.5 – Flash Flooding on Risky River.
Note: Contour lines denote inches of rainfall.

25. What three cities were located in the areas that experienced the highest precipitation?

    • _____

    • _____

    • _____

26. Based on this map, where would you expect most flood and mass movement damage from the storm to occur (use general direction, e.g., SE of a particular site)._____

    Explain your answer: _____

    _____

    _____

    _____

27. The chart below depicts peak flow at selected sites along Risky River. Compare sites A, B, and C. Do you think the flooding in Lakeside was serious? **Yes** or **No**? (*Hint:* Fill in the chart below to determine the discharge at each area.)

Explain your answer: _____

_____

_____

_____

	Change in Discharge (cfs)	Hours to reach peak flow (hrs)	Discharge(cfs/hr) at peak flow
Site A	465	7	
Site B	11,000	4	
Site C	35,700	1	

28. Provide two primary reasons the flooding was either minimal or extreme near Lakeside.

- _____

- _____

29. Do the peak flow data above support the answer you chose for #26 indicating the direction of greatest runoff and potential flood damage?

Explain your answer:_____

_____

_____

_____

_____

30. Site C is 8.1 miles west of Site B. The flood crest reached Site C at Crushing Canyon 35 minutes after the crest left Site B. What was the average speed (or velocity) of the flood crest in:

a. Miles/hr? _____

b. Ft/sec? _____

Part D: Dynamic Planet

Throughout history, many geographic boundaries, both national and international, have been drawn using rivers to define the borders. Examples include but are not limited to, the Rhine River of Europe, which defines the border between six countries; The Red River, Missouri River, and Mississippi Rivers in the US, each of which serve as borders between various states; and the Rio Grande, which delineates the border between the United States and Mexico.

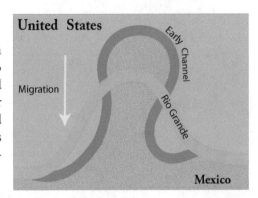

Figure 9.6 – Southern migration of river channel over time

As has been demonstrated, rivers are prone to constant migration, and therefore are not permanent borders. This being the case, many places whose geographic boundaries are defined by rivers, have ended up in boundary disputes with their neighbors. The exercise below is designed to help you understand the nature of these disputes, such that you can make an educated assessment of whether rivers make sustainable boundaries.

Background

The **Chamizal dispute** was a border conflict over about 600 acres on the U.S.-Mexico border between El Paso, Texas, and Ciudad Juárez, Chihuahua. The U.S.–Mexico border was defined based on the location of the Rio Grande in 1852. Within a few decades, the river path had substantially altered resulting in a dispute between the United States and Mexico that was not officially resolved until 1963.

The Treaty of Guadalupe Hidalgo (which officially ended the Mexican–American War) and the Treaty of 1884 were the agreements originally responsible for the settlement of the international border, both of which specified that the middle of Rio Grande was the border– irrespective of any **natural, gradual** alterations in the channels or banks. This provision followed the long–established doctrine of international law that when changes in the course of a boundary river are caused by *natural, gradual* changes, the boundary changes with the river, but when changes are due to *sudden, catastrophic* circumstances, the old channel remains the boundary.

The river continually shifted south between 1852 and 1868, with the most radical shift in the river occurring after a flood in 1864 (see Figure 9.6).

31. Before you examine the details surrounding this conflict, consider the flood of 1864. Based on the international law, should the border change with the new path, remain with the old channel boundary, or is there a case to be made for both sides? Explain _____

_____

_____

_____

By 1873 the river had shifted approximately 600 acres from its earlier position. The newly exposed land came to be known as *El Chamizal*. Both Mexico and the United States claimed the land.

32. Figure 9.7 is a map showing the migration of the river from 1852 to 1963 when a settlement was finally reached. Using a red pen, trace the boundary of *El Chamizal*—the area that prior to 1852 belonged to one country and after decades of river migration, was then considered part of the other country.

    a. Based exclusively on the river channel location, which country did *El Chamizal* belong to prior to 1852? _____

    b. Based exclusively on the river channel location, which country did *El Chamizal* belong to following 1873? _____

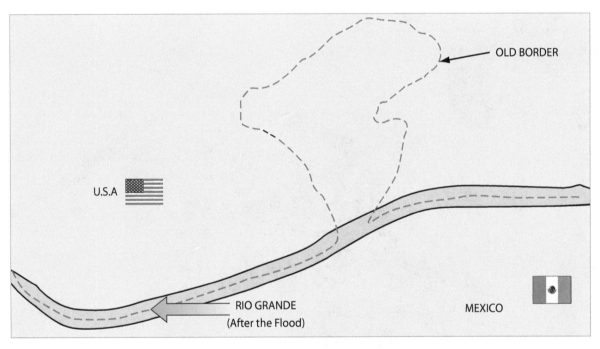

Figure 9.7 – Changes in river path from 1852–1963

33. Based on your answer to the questions above, do you think Mexico claimed the flood was catastrophic or natural? _____ How about the US? _____

    Explain. _____

    _____

    _____

In 1895, Mexican citizens filed suit in the Juárez Primary Court of Claims to reclaim the land. The dispute was formally settled on 14 January 1963. The agreement awarded to Mexico 366 acres of the Chamizal area and 71 acres east of the adjacent Cordova Island. Although no payments were made between the two governments, the United States received compensation from a private Mexican bank for 382 structures included in the transfer. The United States also received 193 acres of Cordova Island from Mexico. On 17 September 1963, the U.S. Congress introduced the American–Mexican Chamizal Convention Act of 1964, which finally settled the matter. In October 1967, President Johnson met with President Gustavo Díaz Ordaz on the border and formally proclaimed the settlement.

34. Examine figure 9.8. Note that the river has been "relocated." This part of the channel is man-made and fortified with concrete. The two governments shared the cost of stabilizing this part of the channel. Why would the governments have cooperated to do this? _____

_____

_____

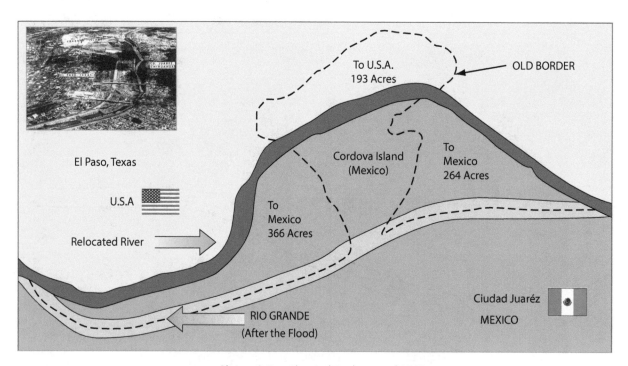

Figure 9.8 – Chamizal Settlement of 1963

Part E: Tying it all Together—Flooding and Its Impacts

Red River Flooding, North Dakota, March 2009

A chart showing known and predicted river response for the Red River as of 7:00 AM on March 27, 2009 is provided below (Figure 9.9). Use this figure to answer the following questions.

35. What is the designated flood stage for the Red River at Fargo, ND? _____ feet.

36. The charts were printed out on March 27, 2009. What was the river stage on March 27, 2009? _____ feet.

37. How many feet above flood stage was the river on March 27, 2009? _____

38. Based on the chart, was the river stage rising or falling on March 27, 2009? _____
    Is the river stage predicted to be higher or lower on March 28, 2009? _____

39. What was the predicted peak flood stage as of March 27, 2009? _____

40. Provide the date and time of the anticipated peak discharge for the Red River at Fargo, ND.

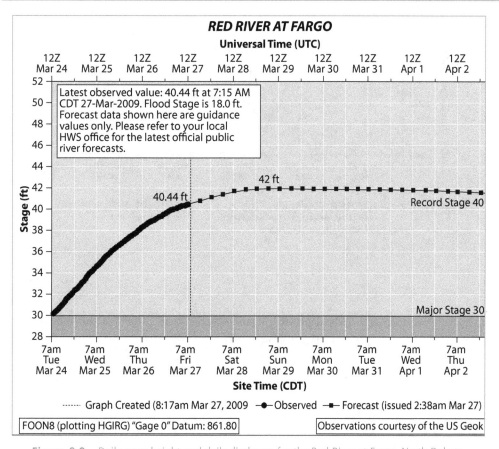

Figure 9.9 – Daily gage height and daily discharge for the Red River at Fargo, North Dakota.

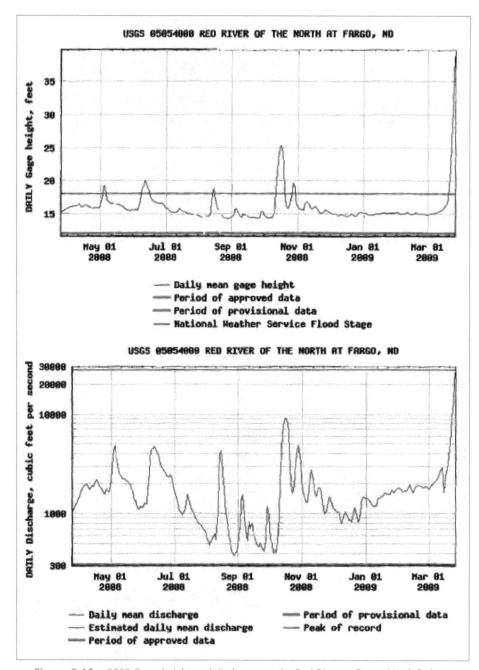

Figure 9.10 – 2009 Gage height and discharge on the Red River at Fargo, North Dakota

Use Figure 9.10, which depicts daily gage height and discharge in 2009 for the Red River at Fargo North Dakota, to answer the questions that follow.

41. Note that the chart showing discharge uses a logarithmic scale to depict discharge. What was the approximate daily discharge of the Red River on March 27, 2009? _____
What was the approximate discharge for the Red River at Fargo, ND on March 1, 2009?

_____

42. By what percentage did the discharge increase between March 1 and March 27, 2009? _____
    Show your work.

43. Using the list of potential flood impacts below established for Fargo by the USGS in conjunction with City personnel, identify with the letter "A" flood impacts below that were likely observed between March 24, 2009 and 7 PM on March 25, 2009 (refer back to Figure 9.9). Identify with the letter "B" flood impacts that might have been observed between March 26, 2009 and March 30, 2009.

A/B	Flood Stage	Impacts
	25.0	Parks and recreation areas along river begin to flood.
	28.0	12th Street toll bridge is closed; begin construction on sewage plant if river is forecast to reach 34 feet.
	30.0	Flooding at 2nd and 3rd Avenue North areas; 2nd Street closed between Main and 4th.
	31.0	1st Avenue North underpass is closed.
	32.5	1st Avenue Bridge across Red River.
	38.2	Covenant Bridge at 52nd Avenue is closed.
	40.0	River is lapping at the base of the Heritage Hjemkomst Interpretive Center.
	40.3	River begins to spill over emergency levees.
	41.4	River begins to spill over Island Park permanent levee.

44. As a city planner, at what flood stage (based on the above) might you begin to order evacuations of select areas? _____ ft

    Explain your answer: _____

    _____

    _____

    _____

Part F: Thinking About Flooding and Flood Control

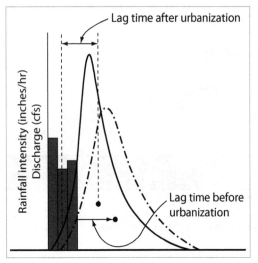

Figure 9.11 – Effect of decreased lag time on flood peak. DATA FROM DUNN AND LEOPOLD, 1978.

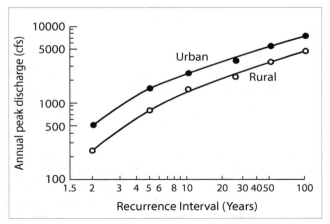

Figure 9.12 –Change in annual peak discharge for various frequencies when area changes from rural to urban. DATA FROM DUNN AND LEOPOLD, 1978.

45. Examine Figures 9.11 and 9.12 to answer the following questions.

   a. Does urbanization typically increase or decrease flooding? _____

   b. How does the recurrence interval change when a rural water shed is urbanized (i.e.—does a flood of a given magnitude happen more frequently, less frequently, or doesn't change, when a rural area is urbanized)? _____

   c. How does the lag time change for a given stream with respect to urbanization of a rural stream? _____

   d. What 2 main features, inherent in urbanization likely causes the changes you specified in questions b and c?

   • _____

   • _____

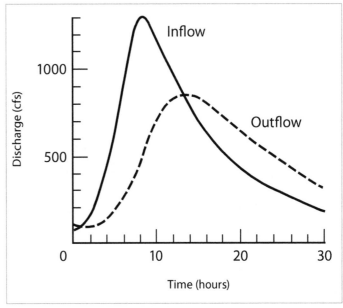

Figure 9.13 – Hydrograph of inflow to a reservoir, solid line; and of outflow from reservoir, dashed line. DATA FROM DUNN AND LEOPOLD, 1978.

46. Examine Figure 9.13 which shows a hydrograph for a storm water retention basin. Use this figure to answer the questions that follow:

a. What is the purpose of a storm water retention basin? i.e., what specifically does the storm water basin do to aid in the prevention of flooding? (Think about peak discharge and recovery time).

_____

_____

_____

_____

b. List three natural factors (i.e., not related to urbanization) that affect the degree to which flooding does or does not occur as a result of any given amount of rainfall (amount of precipitation does not count!). In order to answer this question, think about the climate and soil conditions.

    • _____

    • _____

    • _____

c. List and describe three different ways used to mitigate flood damage.

    • _____

    • _____

    • _____

GROUNDWATER HYDROLOGY

INTRODUCTION

The source of all fresh water on the planet is atmospheric precipitation. As part of the hydrologic cycle, the atmosphere evaporates salt water from the ocean and rains in back down onto the planet as fresh water, which then will either: 1) become incorporated into the ice caps; 2) join the river systems; 3) be consumed by vegetation; or 4) filter into the ground and replenish the groundwater reservoir.

Groundwater is water contained below the surface within saturated sediment and rock. It arrives there by percolating into the ground and navigating downward through open spaces in the substrate. The area where spaces within the substrate are primarily occupied by air serves as a corridor through which the water must pass before reaching the groundwater reservoir. This relatively dry part of the subsurface is called the "zone of aeration" or the "unsaturated zone," as the openings are not fully filled with water. Due to the influence of gravity, the water will continue its downward descent until compaction of the sediment or the presence of an impermeable layer prevents additional movement. At this point, the water will pool and fill in open spaces between sediment grains and fractures within rock, generating an area known as the "zone of saturation." Water in this zone is called "groundwater" and the upper limit of this zone, which divides the zone of aeration from the zone of saturation, is called the "watertable." See Figure 10.1.

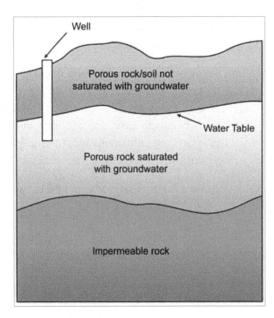

Figure 10.1 – Groundwater System

Groundwater moves slowly down-gradient through pore spaces between the sediment until it reaches a point where the surface intersects the water table. The groundwater is then able to discharge to the surface in the form of a stream, spring, lake, or wetland, or be withdrawn through a well. Once brought to the surface, it then rejoins the surface hydrologic cycle and will eventually be evaporated and subsequently returned to the atmosphere, from which it will be rained out again. The groundwater we use today may have traveled through the hydrologic cycle hundreds of thousands of times since Earth was formed.

Understanding where the groundwater reservoirs are located, how and what rate it's transported, what techniques are available to bring it up to the surface, and how to responsibly use and preserve this precious resource is critically important - as groundwater is the primary supply of fresh water for human consumption.

OBJECTIVE

The objective of this lab is four-fold: 1) Explore where groundwater is stored; 2) Develop an understanding of the factors that govern groundwater transport; 3) Examine the relationship between the direction and rate of groundwater flow and the location of water wells 4) Incorporate this knowledge while investigating what enhances and retards groundwater contamination.

Part A: Porosity, Permeability, and Aquifers

Groundwater is housed and transported through "pores," which are fractures within rocks and between grains of unconsolidated material. "Porosity" is the volume of void spaces within and between the geologic material and is expressed as a percentage.   While porosity defines the total amount of water that can be housed, not all the water in the pore is available for transport. When groundwater is subjected to the influence of gravity, most of the water is set in motion, but molecular forces and surface tension will retain some of the water in the pore. The volume of water that can be moved is called "specific yield" while the quantity of water retained in the pore is called "specific retention." Specific yield plus specific retention equals porosity; i.e. the total volume of water than can be housed in the voids.

"Permeability"- aka "hydraulic conductivity"- is a measure of the ease with which water moves from one pore to another. This, in turn is dependent on the quantity, size, and sinuosity of the conduits that connect the pores. Wider and straighter conduits promote higher permeability—while a higher quantity of conduits increases the amount of water that can be transported. A body of rock or sediment that transports groundwater easily is called an aquifer, one that grudgingly transports groundwater is called an aquitard, and a rock body incapable of transporting groundwater is called an aquiclude.

Aquifers are divided into two basic types (see Figure 10.2):

A.   Confined Aquifer - Aquifer that is separated from the surface by an "aquitard" - sediments or rocks that do NOT transport water easily and therefore retard the motion of water (ex: mudstone or claystone)

B.   Unconfined Aquifer – Aquifer in which there is no confining layer between the water table and the surface. – usually composed of "Alluvium" - the name given to sediments that have been weathered and eroded by some form of water then deposited in a non-marine setting. Typically made of sand and gravel.

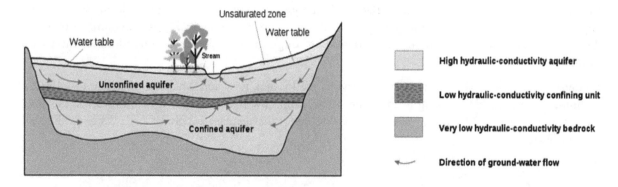

Figure 10.2 – Basic Types of Aquifers

Bearing in mind that the ability of a geologic substrate to be a good aquifer is dependent primarily on its permeability, it's important to understand that there is a distinct difference between a material that is "porous" and a material that is "permeable."

Water

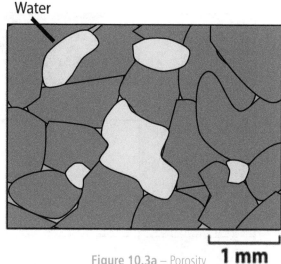

– Porosity — The percentage of open pore spaces (empty space) in a given volume of rock or sediment that determines the total amount of water a material will hold. Porosity is largely influenced by particle size, shape, sorting, and compaction. See Figure 10.3a.

Figure 10.3a – Porosity   **1 mm**

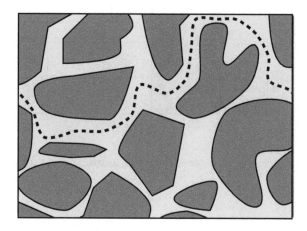

– Permeability — The ability of a material to allow fluids to pass through an interconnected network of pores. See Figure 10.3b.

Figure 10.3b – Permeability

A material can be very porous, but if the pores are not interconnected, groundwater cannot flow through material, and the material will therefore be a poor aquifer. As such, in order for a material to make a good aquifer, it must be porous AND permeable. Porous materials that are not permeable are not aquifers.

Exercise: What substrate makes the best aquifer?

Table 10.A depicts a list of common geologic materials and their porosity and permeability.

Table 10.A

	Geologic Material	Hydraulic Conductivity (lower limit) (m/yr)	Hydraulic Conductivity (upper limit) (m/yr)	Average Hydraulic Conductivity (m/yr)	Average % Porosity
Unconsolidated Sediment	Well Sorted Gravel	10000	10000000		37.5
	Well Sorted Sand	100	100000		37.5
	Silty Sand	10	10000		15
	Silt	0.01	100		42.5
	Glacial Till	0.00001	10		15
	Clay	0.00001	0.01		50
Consolidated Rock	Unfractured Igneous and Meta Rock	0.0000001	0.0001		1
	Shale	0.000001	0.01		5
	Sandstone	0.0001	10		29
	Fresh Limestone	0.01	10		5
	Karst Limestone	10	100000		n/a

Refer to Table 10.A when answering the following questions:

1. Calculate the average hydraulic conductivity for the geologic materials listed in Table 10.A and place your answers in the column provided.

2. Considering only the unconsolidated samples that contain some quantity of med and/or coarse grains (i.e., gravel, sand, silty sand, and glacial till), based on the above data, which type of sediment makes a better aquifer – sorted or unsorted? _____. Explain why this is the case._____

    _____

    _____

    _____

3. Using the average hydraulic conductivities, calculate how much more efficient Karst limestone is at transporting groundwater than fresh, unweathered limestone. _____x. Explain why this is the case _____

    _____

    _____

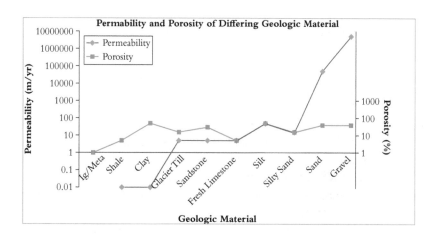

Graph 10.I – Porosity and Permeability of Common Surface Materials

4. Graph 10.I depicts the porosity and permeability of some of the most common surface materials. Using that graph, answer the following questions.

    a. Is there any clearly quantifiable relationship between porosity and permeability of a given geologic material? Yes or No?

    b. Based on the graph and your answer to the question above, does a higher porosity necessarily mean higher permeability? Yes or No?

5. Given that there are infinite varieties of sediment and rock, each of which depicts different properties from the next, if you were a consulting geologist in search of a lucrative aquifer to supply a growing city, what is the primary property you would evaluate in the subsurface geology?

_____

Part B: Investigating Sediment Permeability

A. Obtain a clean soil tube and make sure the cap is securely on the bottom.

B. Construct a layered column of sediment as shown in Figure 10.4

C. Mix a few drops of food coloring with some water and slowly pour the water into the tube. Observe how the water flows through each layer of your column

1. Which sediment type has the greatest potential for transporting pollution?
    _____

    Why? _____

2. Which sediment type has the least potential for transporting pollution?
    _____

    Why? _____

3. Based on the experiments you just performed, explain why clay is frequently used in lining landfills.

Figure 10.4

Part C: Getting Water to the Well and up to the Surface

Understanding Groundwater Transport

The velocity of groundwater transport is calculated by multiplying the permeability of the material by the hydraulic gradient then dividing by the specific yield. (Note—When groundwater is set in motion, a small quantity of water remains in the pore due to molecular forces and surface tension. Specific yield is the volume of water that gets removed from the pore—i.e., the amount of water in pore minus that small bit that got retained.) Often times, the permeability (p) is expressed in units such as ft/day or cm/sec). The hydraulic gradient (g) is the difference in water level per unit of distance in a given direction (e.g., the slope of the water table). Just like a slope on the surface, the hydraulic gradient can be measured directly from water-level maps in ft/ft or ft/mile. The following formula will allow you to calculate groundwater velocity in ft/day, where the specific yield (y) is given as a decimal (i.e., 10% = .10) and (g) is in ft/ft.

$$v = \frac{p * g}{y}$$

With respect to the direction of flow, as a general rule, the water table forms a gently sloping surface that mimics the land surface. As such, the water table can be mapped and contoured just like surface topography. In this case, however, control points are water elevations in wells, springs, lakes, or streams, as opposed to surface elevations. The direction of groundwater flow can be indicated by flow lines, which are drawn perpendicular to water-level contour lines in the direction of flow. See figure 10.5 for visual aid.

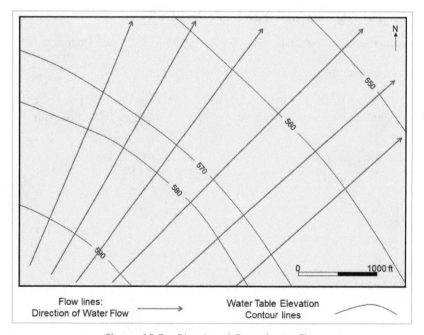

Figure 10.5 – Direction of Groundwater Flow

Groundwater moves slowly down gradient through the pore spaces between the sediment until it reaches a point where the surface intersects the water table. There it can naturally discharge to the surface in the form of a stream, spring, lake, or wetland, while new atmospheric precipitation seeps into the ground at locations known as recharge sites and replenishes the groundwater lost to the surface. However, groundwater can also be prematurely withdrawn from the subsurface and brought up for human consumption through manmade devices called wells.

Wells are holes that people dig or drill such that water can be withdrawn as needed at a rate controlled by the demands of the population. The wells are designed to penetrate through the zone of aeration and into the zone of saturation—so water can be continuously withdrawn as long as the water table remains higher than the base of the well. Wells are classified into two basic categories: Ordinary and Artesian (see Figure 10.6).

Ordinary wells are constructed by digging a hole into the aquifer that infiltrates deeper than the water table. Water flowing through the aquifer then permeates the well and fills it—such that the water surface in the well is even with the water table. In order get water from the well to the surface, it must be manually withdrawn, which can be done with anything from buckets to state of the art pumps. The amount of water artificially discharged (withdrawn) from a well of this type must be carefully monitored to ensure there is adequate recharge replacing it, or the water table will drop and eventually cause the well to go dry, in addition to inducing a plethora of environmental problems.

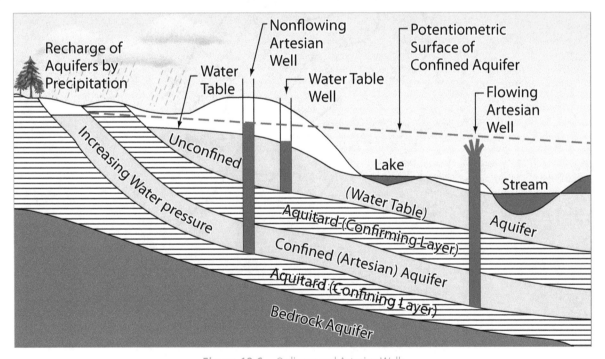

Figure 10.6 – Ordinary and Artesian Wells

Artesian wells differ from ordinary wells in that water is able to rise at least part ways to the surface without any pumping or other forms of manual withdrawal. Artesian wells require very specific geologic conditions in which an aquifer lies beneath an aquitard and both are sloping away from the recharge area. A setup of this type puts the water in the aquifer under tremendous pressure and the pressure alone will push water up through any vertical opening, including a manmade well. In some cases, the pressure is sufficient to drive the water all the way to the surface, where it generates a "flowing artesian well." In other cases, the water is not pushed high enough to reach the surface and results in "non-flowing" artesian wells. Which one develops is dependent on the location of the "potentiometric surface"—an imaginary line where a given reservoir of fluid will equilibrate if allowed to flow. Once the ascending water reaches the potentiometric surface, the pressure equalizes and the water will go no higher.

Exercise: Ordinary Wells and the Water Table

Figure 10.7 depicts the topography of an area characterized by flowing surface water, which is underlain by a shallow, unconfined aquifer.   Use Figure 10.7 to answer the following questions.

Figure 10.7 and corresponding graph

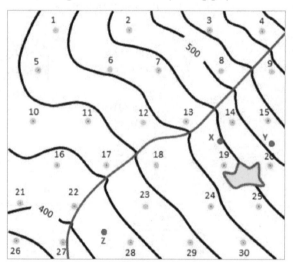

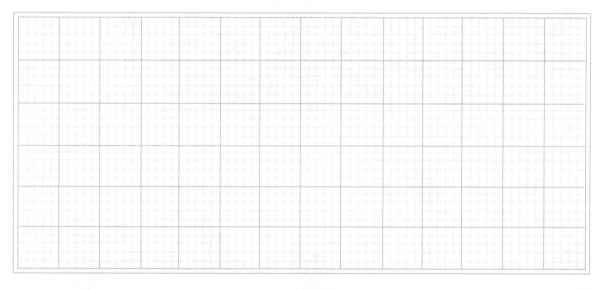

1. On the graph provided, construct a topographic profile of figure 10.7 from the NE corner to the SW corner of the map.

2. Shallows holes were drilled throughout the landscape in order to determine the depth of the water table below the surface. The locations that were sampled are labeled 1–30 on fig 10.7 and the depth to the water table for each location is listed in Table 10.B. For each of the sampled locations, calculate the water table elevation in feet above sea level by following the procedure below:

    a. Determine the surface elevation at samples sites 1-15 on fig 10.7. (note: record the surface elevation using 5 ft contour measurements—i.e. 455, 460, 465, etc.). Sample sites 16–30 have been filled in for you.

    b. Subtract the water depth from the surface elevation for all the sample sites to calculate the water table elevation above sea level.

Record your answers for both the surface elevation and the elevation of the water above sea level in the spaces provided in Table 10.B.

Table10.B – Sample Site Water Level Data

Site #	Surface Elevation	Depth to Water Table (Ft)	Water Table Elevation (ft above sea level)	Site #	Surface Elevation	Depth to Water Table (Ft)	Water Table Elevation (ft above sea level)
1	455	16	439	16	415	13	
2		14		17	425	14	
3		21		18	445	16	
4		24		19	470	18	
5		15		20	495	20	
6		16		21	405	12	
7		18		22	410	12	
8		20		23	430	14	
9		23		24	455	17	
10		15		25	475	19	
11		14		26	375	25	
12		17		27	390	21	
13		17		28	415	18	
14		20		29	435	15	
15		22		30	455	18	

3. Next, transfer the water table elevations you just calculated onto Figure 10.7 such that each sample site depicts the elevation of the water table in that location.

4. Using a red pen, contour the water elevation lines in 20 ft intervals, beginning with 350.

5. On the same graph paper you used in #1 above, and using the same scale as before, construct a second topographic profile, from the NE corner to SW corner of fig 10.7, this time using the water table contours.

6. What is the relationship between the elevation of the water table and the topography of the surface landscape? _____. Explain why this is the case? _____
_____
_____

7. Remember that groundwater flow occurs at right angles to the water elevation contour lines. Keeping this in mind, draw arrows on fig 10.7 to indicate the direction of groundwater flow in the aquifer.

8. Calculate the hydraulic gradient between the NE and SW corners of the map in ft/ft.—i.e., the average hydraulic gradient for the area in question. _____ft/ft

9. Using the equation below and the values in Table 10.C, calculate the groundwater velocity in each of the 4 substrates represented in figure 10.8. (Note—the average hydraulic gradient is not sufficient for this calculation. You will need to calculate the hydraulic gradient for each individual area comprised of a different sediment before you can calculate the groundwater velocity.) Record your answers in spaces provided in Table 10.C.

$$V = \frac{(\text{Hydraulic conductivity (ft/day) x hydraulic gradient (ft/ft)})}{\text{Specific Yield}}$$

Table 10.C – Effect of Sediment Type on Groundwater Movement and Yield

Sediment Type	Hydraulic Conductivity (ft/yr)	Hydraulic Gradient (ft/ft)	Specified Yield (%)	Groundwater Velocity (ft/day)
Gravel	219,000		22	
Sandy Gravel	365		25	
Coarse Sand	146,000		27	
Fine Sand	3,650		21	

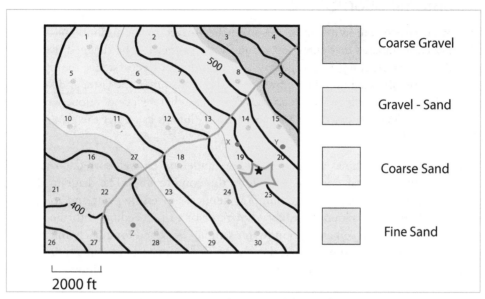

Figure 10.8 – Geologic Map of the Aquifer

10. The yellow pond in figure 10.8 marks the boundary of a landfill that has been leaking contaminants into the groundwater system from a spot marked by the black star. There are 3 wells throughout this area that are used by surrounding communities for their municipal water supply. Which of the 3 wells is at risk for being contaminated: X,Y, or Z? (circle one)

11. Based on your answers to question 9, approximately how long will it take the contaminant to get from the contamination site to the well you chose in question 10? *Show your work*
    _____ days

Part D: Contamination Scenario

Remember that because groundwater infiltrates from the surface to the ground and then back to the surface again, any contaminants that enter the groundwater system can be dragged over large distances for long periods of time, resulting in the destruction of numerous ecosystems, multiple public health problems and ruination of the aquifer. The more permeable the aquifer, the faster and farther the pollutant can spread. This is particularly pertinent in today's world, where every developed area has industrial waste, landfills, sewers, and countless other ways in which pollutants can enter the groundwater system if not carefully controlled.

Figure 10.9 is a diagram that shows a hypothetical area underlain by well-sorted, coarse sand ~30 feet thick. The depth to the water table is indicated along the contour lines. The diagram shows three potential sources of pollution: an open garbage dump, a chemical processing plant, and an acid mine dump. The site for a new housing development is also shown. Prior to proceeding with the project, the land developer ordered that random soil and groundwater samples be tested for any chemical contamination above the legal limits. Test results showed that the area labeled "initial site of contamination" was heavily contaminated by one chemical. The developer hired an environmental consulting firm to evaluate the initial findings further.

In this activity, you will act as the consulting firm and try to determine the extent of the contamination plumes for each possible source, as well as isolate the single source of contamination to the housing development site.

1. You have been given a budget to drill a well at each of the 28 well-monitoring sites in order to test the groundwater and conclusively identify the pollution source to the development site. To do this, follow the instructions below.

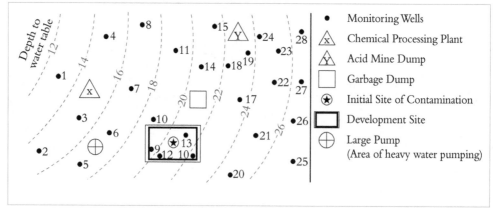

Figure 10.9 – Diagram of Contaminated Site and Surrounding Areas

a. On the counter, you will find sample water from each of the 28 monitoring wells in the diagram above.

b. Using the designated pipette, put a drop of solution from the first well sample onto a piece of litmus paper. Observe the color change and record your results in Data Table 10.D.

c. Repeat the procedure for the other 27 well samples.

d. Interpret your results using the key provided beneath the data table and record your data in the chart.

Data Table 10.D

Monitoring Well Sites	Color	Concentration (ppm/ppb)
1		
2		
3		
4		
5		
6		
7		
8		
9		
10		
11		
12		
13		
14		
15		
16		
17		
18		
19		
20		
21		
22		
23		
24		
25		
26		
27		
28		

Key

Yellow–Orange color = "No pollution" < 1 ppm
Light–Dark Green color = "Low pollution" < 3 ppm
Red–Brown color = "High pollution" >5 ppm

Note: Assume that the legal allowable limit for this particular pollutant is 1 ppm.

2. Based on the information provided in Figure 10.9, draw in 3 flow lines to indicate the direction of groundwater flow. One at the top, one in the middle, and one at the bottom of the diagram. (*Hint:* Flow lines will always be at right angles to the contour lines with the arrow pointing in the direction of lower elevation.)

3. Transfer your data from Data Table 10.D to Figure 10.9 by putting a "0" by the sites that showed no pollution, an "L" by the sites that showed low pollution, and an "H" by the sights that showed high pollution.

4. Based on the direction of groundwater flow and the concentration of contaminant at each site, on Figure 10.9, draw the likely contamination plumes for each source of pollution. (*Hint:* The plume would begin at the top and bottom of the contaminant source and get progressively wider as it flows downstream. See Figure 10.10.)

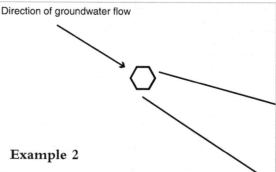

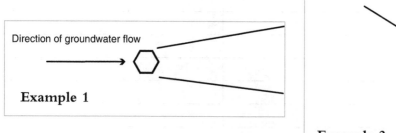

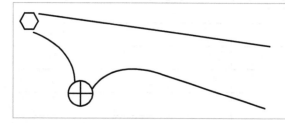

WARNING: If there is heavy pumping in the area, the well will suck in the contaminant, so keep that in mind when drawing your plumes.

Figure 10.10 – Path of Contamination Plumes

Remember that any well sites within a contamination plume will have a high concentration of pollution; those close to, but not included in the plume will have a low concentration of pollution; and the wells far away from the plume will have no pollution.

5. Based on your drawn contamination plumes, which of the 3 potential sources of pollution is responsible for the pollution problems at the housing development site? _____

GROUNDWATER AND POLLUTION

INTRODUCTION

Groundwater is housed and transported through "pores," which are openings within rocks and between grains of unconsolidated material. "Porosity" is the volume of void spaces within and between the geologic material and is expressed as a percentage. When groundwater is subjected to the influence of gravity and pressure, most of the water is set in motion, but molecular forces and surface tension will retain some of the water in the pore. The volume of water that can be removed by gravity drainage is the "specific yield," while the quantity retained is the "specific retention." Specific yield plus specific retention equals porosity; i.e., the total amount of water that can be housed in the voids.

"Permeability" is a measure of the ease with which water moves through geologic material. A body of rock or sediment that transports groundwater easily is called an aquifer, one that grudgingly transports groundwater is called an aquitard, and a rock body incapable of transporting groundwater is called an aquiclude. When groundwater becomes contaminated, the foreign pollutant will be transported wherever the groundwater travels, which can result in ruination of aquifers, contamination of soil and surface waters, and a plethora of additional environmental problems.

As a general rule, the water table forms a gently sloping surface that mimics the land surface. As such, the water table can be mapped and contoured just like surface topography. In this case, however, control points are water elevations in wells, springs, lakes, or streams, as opposed to surface elevations.

OBJECTIVE

The objective of this lab is two-fold: 1) to gain an understanding of the direction and rates of groundwater movement, and 2) to apply that knowledge to various contamination scenarios in an attempt to examine the environmental consequences associated with misuse of groundwater.

Part A: Calculating Groundwater Movement

The velocity of groundwater transport is calculated by multiplying the permeability of the material by the hydraulic gradient, then dividing by the specific yield. Oftentimes, the permeability is expressed in units such as ft/day or cm/sec, but the process of groundwater transport is easier to conceptualize when working with more familiar units. As such, the equation below permits you to calculate groundwater velocity using a permeability value (p) that is expressed in gallons per day per square foot (gpd/ft^2).

The hydraulic gradient (g) is the difference in water level per unit of distance in a given direction (e.g., the slope of the water table). Just like a slope on the surface, the hydraulic gradient can be measured directly from water-level maps in ft/ft or ft/mile. Using the fact that there are 7.48 gal per cubic foot, the following formula will allow you to calculate groundwater velocity in ft/day, where the specific yield (y) is given as a decimal (i.e., 10% = .10) and (g) is in ft/ft.

$$V = \frac{p * g}{7.48y}$$

The direction of groundwater flow can be indicated by flow lines, **which are drawn perpendicular to water-level contour lines in the direction of flow**.

Use Figure 11.1 to answer the following questions. *Show your work.*

1. What is the average water-level gradient along the eastern flow line?

    a. _____ ft/mi

    b. _____ ft/ft

2. If an aquifer located in the vicinity of figure 11.1 is 25 ft thick and has a porosity (specific yield) of 15%, how much water is stored in a .5 mi^2 area? *Show your work.*   _____ ft^3

    Now multiply your answer by 7.48 to convert to gallons: _____ gallons

3.   What is the groundwater velocity in the vicinity of the eastern flow line, given that the permeability of the aquifer is 750 gpd/ft^2 and the specific yield is 20%? *Show your work.*
_____ ft/day

4.   Flow lines are used to depict the direction of groundwater flow **and are drawn at right angles to the contour lines**. Using a red colored pencil, construct a flow line on figure 11.1 from the red dot. In what direction is the water moving? _____

5.   If chemical contaminants from an industrial plant spilled at the site marked by the red dot, which body of water would it contaminate? _____

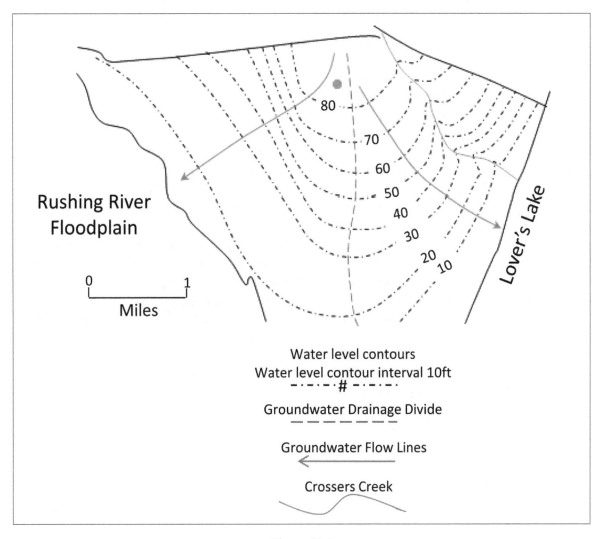

Figure 11.1

Part B: Groundwater Contamination

Groundwater may be contaminated by improper storage or disposal of wastes at the surface. Unfortunately, contamination in this manner happens all too often due to ignorance about groundwater flow and potential adverse health effects, as well as the misconception that water is a renewable resource and therefore will be plentifully available in the long term. One such example involves the storage and/or disposal of saltwater that is brought to the surface through the mining of oil.

When oil is mined from the ground, brine, aka "saltwater," is produced with the oil; the brine is usually of little or no economic value. As such, when unregulated, it is often disposed of in the most economical manner. Many times, the unwanted brine is pumped into unlined, easily infiltrated holding ponds or pits throughout the terrain. Due to the inadequate containment, most of the waste infiltrates the ground. When this salty brine reaches the water table and mixes with the fresh water, it can lead to severe groundwater pollution, as the chloride concentrations in the brines may exceed 35,000 mg/L. Most areas of uncontaminated groundwater have chloride concentrations of < 25 mg/L. The US EPA recommends that drinking water contain no more than 250 mg/L of chloride, since the resulting salty taste will make the water inadequate for human consumption. In addition to making the water undrinkable, water with higher chloride concentrations will sterilize the soil and kill vegetation.

In this exercise, we look at a hypothetical scenario that is based on an actual occurrence, where an industrial oil company attempted to dispose of liquid waste by pouring the unwanted material into "holding ponds" throughout the area.

The following exercise is based on several studies conducted at a brine-contaminated site on a nearly flat floodplain in the western US. Several oil wells were drilled in this vicinity in 1972. Approximately 10x more brine than oil was brought to the surface and nearly 275,000 barrels of saltwater were pumped into 4 ponds from August 1972 to September 1973. The brine contained ~34,500 mg/L of the chloride ion.

Figures 11.2 and 11.3 show the locations of 4 brine disposal pits, 3 oil wells, 23 observation wells, and a water well. The observation wells were installed in November 1973, at the completion of brine disposal, to monitor contaminated groundwater movement. Shale is less than 30 ft deep and is overlain by deposits consisting of a mixture of sand, silt, and clay. The average permeability (p) of the river sediments, which contains contaminated water, is about 225 gpd/ft^2 and the average specific yield (y) is 0.18. The hydraulic gradient (g) can be determined from the information in Figure 11.2.

Note: Water table elevations from the observation wells in April 1977 were used to draw in water-table contour lines every 2 feet in Figure 11.2.

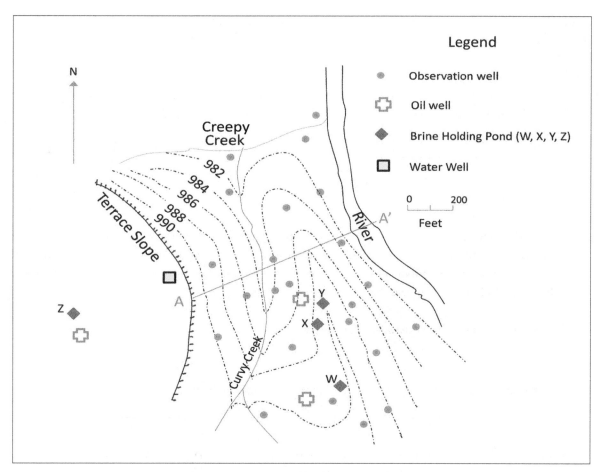

Figure 11.2 – Water table elevations, April 1977.

Use Figure 11.2 to answer the following questions. *Show your work.*

6. Draw several flow lines originating at the brine holding ponds (remember there are 4, labeled W, X, Y, and Z respectively) to the most likely area of groundwater discharge (keep in mind that groundwater will flow at right angles to the contour lines). Note also that water will flow in *all* downslope directions, meaning the water will flow east, west, and north from each of the brine ponds.

7. What is the hydraulic gradient from pond Y to the river? _____ ft/ft

8. What is the hydraulic gradient from pond Y to Creepy Creek? _____ ft/ft

9. The velocity of groundwater from pond Y to the river is _____ ft/day

10. The velocity of groundwater from pond Y to Creepy Creek is _____ ft/day

11. If we divide the distance of travel measured along the flow line by the rate of groundwater flow, we obtain the travel time (d = rt). What are the travel times (in days) for water from pond Y to:

a. River _____ days (*show your work*)

b. Creepy Creek _____ days (*show your work*)

Note: Chloride concentrations at the various observation wells in December 1974 were used to draw the map in Figure 11.3. Each contour line represents an area of equal chloride concentration and therefore shows the distribution of contaminant in this area during December 1974. There is a change of 4000 mg/L between two adjacent contour lines.

Use Figure 11.3 to answer the questions that follow:

12.

a) Examine Figure 11.3. Where are the highest concentrations of chloride located?

_____

b) Do the highest concentrations of chloride on Figure 11.3 correspond with the locations of your flowlines on Figure 11.2? **Yes** or **No**? _____

c) Does this make sense? Explain. _____

_____

_____

13. Are the surface waters also at risk as a result of this contamination? **Yes** or **No**? _____
Explain: _____

_____

_____

14. The two purple observation wells in figure 11.3 were singled out because they contained higher concentrations of chloride in December 1974 than in November 1973, while all the other wells contained less. Why is this the case?

_____

_____

_____

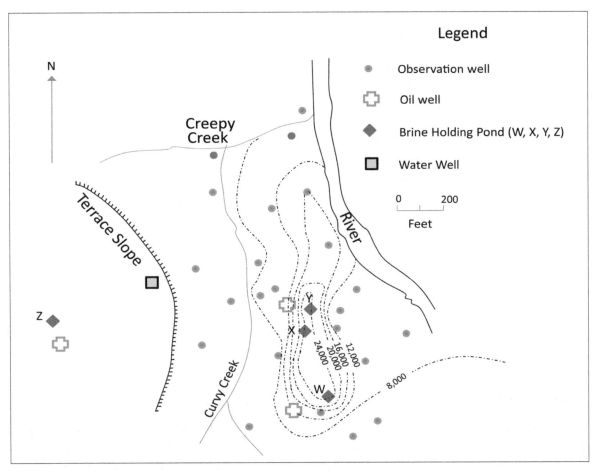

Figure 11.3 – Chloride Concentrations in December 1974.

15. Chlorine is used in the treatment of wastewater. Why does it not contaminate the environment into which the treated water is released?

_____

_____

_____

16. The shallow farm well (11 ft deep) marked by this symbol ☐ on figures 11.2 and 11.3 increased in chloride concentration between 1974 and 1977. Has this contamination resulted from brine disposal into ponds **W, X, Y,** or **Z** (circle one)? Explain with the aid of the cross-section below, which shows you the landforms from A → A′ in Figure 11.2:

_____

_____

_____

_____

Cross-Section of A - A' Line in Figure 11.2

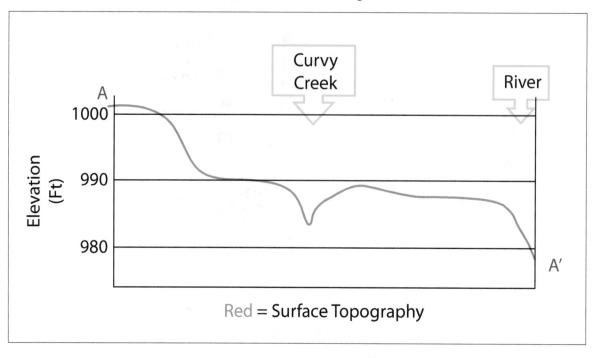

Red = Surface Topography

Part C: Groundwater Overdraft and Flow Reversal

Over-pumping of groundwater reservoirs can have numerous negative effects. Among them are drying out the well, sediment compaction which leads to sinkholes, groundwater flow reversal, and saltwater intrusion. Because much of the groundwater is pumped out for irrigational purposes, drying out the wells can have devastating effects on the economy. Below is an exercise that demonstrates the effects of over-pumping groundwater reservoirs.

The original landscape depicted by figures 11.4 and 11.5 is a primarily flat region, which in conjunction with plentiful water for irrigation and humid climate, provides an ideal setting for agricultural fields. Crops have been grown in this area since 1910. The configuration of the water level in 1920 is shown in Figure 11.4. While the water supply was originally capable of supporting the area's agricultural industry, continuous, intense pumping of water for irrigation from the underlying aquifer has caused a substantial overdraft in the groundwater supply and an equally sizeable decline in the area's water levels.

Banking institutions have made loans to farmers in this area for many years. Because groundwater supplies play such an important role in crop production, the banks require yearly reports on the availability of groundwater. As the water level declines, the cost of pumping the water increases. In the long run, pumping costs when combined with other farm operation costs could be greater than the value of the crop. The economic impact is obvious.

In order to evaluate the extent of water decline and to determine remedial measures, maps of conditions in 1920 and 1960 were prepared and evaluated.

Use figures 11.4 and 11.5 to answer the following questions.

17. Compare Figures 11.4 and 11.5. What are the major differences with respect to:

- Water table elevation _____

- Gradient _____

- Direction of water flow _____

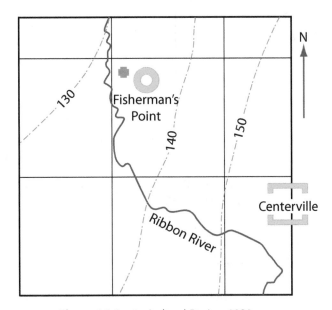

Figure 11.4 – Agricultural Region, 1920.

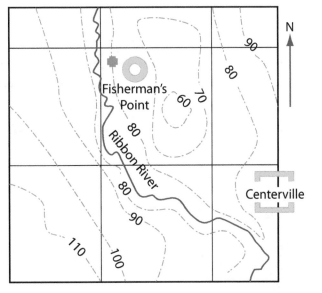

Note: Contour interval is 10 ft. Contour lines mark feet above sea level.

▦ Fish Market

Figure 11.5 – Agricultural Region, 1960.

18. Figure 11.6 shows the net decline in water levels from 1920 to 1960. What area had the greatest decline in water levels? _____

_____

19. Starting at the west edge of Centerville, draw a flowline across the 1920 map (Figure 11.4). Draw a similar flow line passing through Fisherman's Point. What is the general direction of groundwater movement in 1920 at:

a. Centerville

_____

b. Fisherman's Point

_____

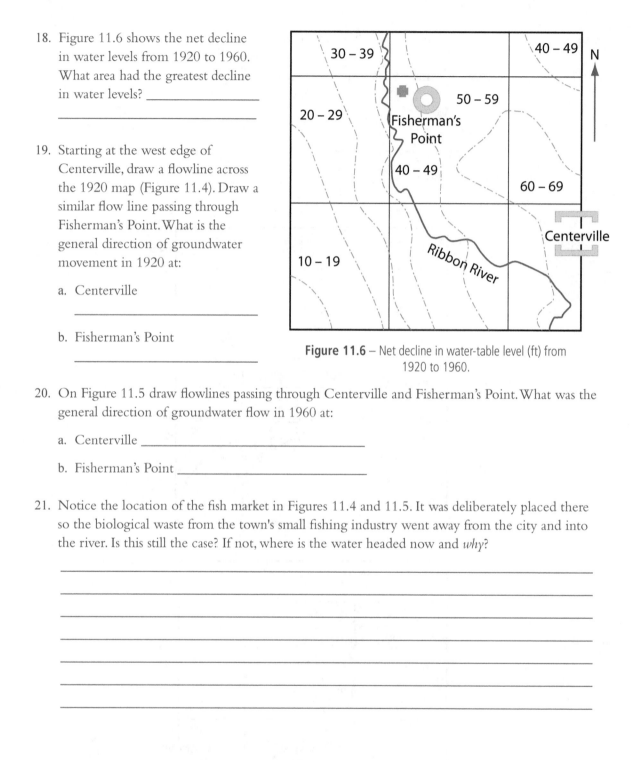

Figure 11.6 – Net decline in water-table level (ft) from 1920 to 1960.

20. On Figure 11.5 draw flowlines passing through Centerville and Fisherman's Point. What was the general direction of groundwater flow in 1960 at:

a. Centerville _____

b. Fisherman's Point _____

21. Notice the location of the fish market in Figures 11.4 and 11.5. It was deliberately placed there so the biological waste from the town's small fishing industry went away from the city and into the river. Is this still the case? If not, where is the water headed now and *why*?

_____

_____

_____

_____

_____

_____

_____

Part D: Saltwater Intrusion

In any city with a significant residential and industrial sector, large quantities of groundwater are used for industrial, municipal, and domestic purposes. As a metropolis expands, more and more water becomes necessary to support the region and this puts a considerable strain on the available water resources in the area. While significant water-level declines in an increasing number of regions, both on the coast as well as inland, have caused concern that the local water supply might become seriously depleted or contaminated, coastal municipalities are particularly at risk, as they face the additional hazard of saltwater intrusion.

Because coastal aquifers are adjacent to the ocean, over-pumping of groundwater in these areas will initially cause deterioration of the water quality by increasing the salinity, but as the pumping increases with time, saltwater may eventually reach the pumping center, contaminate the water supply, and in effect, permanently ruin the aquifer. The following exercise is a hypothetical scenario based on an actual occurrence that demonstrates the threat of saltwater intrusion.

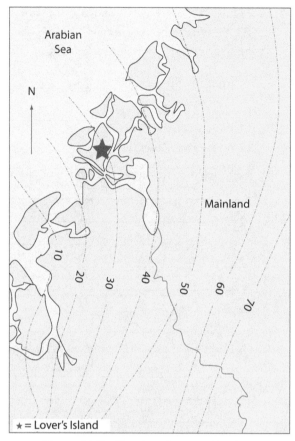

Figure 11.7 – Elevation of water table above sea level, 1900

Use Figure 11.7 to answer the following questions:

22. A water-level map representing conditions that existed in 1900 is shown in Figure 11.7. Construct 4 equally spaced flowlines showing the direction of groundwater movement in 1900.

    a. What was the general direction of groundwater flow in 1900? _____

    b. Was groundwater at Lover's Island likely to have been salty in 1900? **Yes** or **No?** Why or why not? _____

    _____

    _____

    _____

    _____

    _____

23. Figure 11.8 is a water-level map that shows conditions in the same area during 1985. Starting at the SE corner, NE corner, and Lover's Island, draw flowlines on Figure 11.8 showing the general direction of groundwater movement in 1985. In what general direction was the water moving:

a. from the SE corner? _____

b. from the NE corner? _____

c. from Lover's Island? _____

24. From what direction do you expect the fresh- and saltwater interface to first reach the area? _____

_____

Why? _____

_____

_____

_____

25. Is Lover's Island still self-sufficient with respect to their water supply? Yes or no—explain. _____

_____

_____

_____

_____

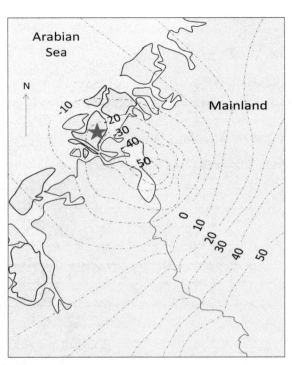

Figure 11.8 – Elevation of water table above
sea level, 1985

WATER POLLUTION AND WASTE WATER TREATMENT

INTRODUCTION

Waste water from domestic and industrial sources must be treated before being released back into the environment. The necessity of this cleansing process is readily apparent with a quick trip around your house during any average day of the week. Beginning with the kitchen, the sink drain is a conduit for fruits, vegetables, grease, soap, and various types of bacteria, all of which go straight to the sewers, along with the food- and soap-laden water produced by the dishwasher. It is also joined by the detergent-filled water from the washing machine, shower and bath water full of hair, skin, bacteria, soap, and shampoo, in addition to water from the toilet, which not only contains bodily fluids, but also any medications or drugs that were in your system at the time of its use. Cuts are frequently washed in the sink, so blood can be added to the mix, as can cleaning chemicals, like Ajax and Comet, in addition to materials from hobbies such as paint and glue. The above are just common household contaminants and are significantly compounded by the pollutants present in public utilities and industrial facilities. As such, all of the materials discussed above are potentially harmful to us as well as the ecosystem. Therefore sewage water must be cleaned prior rejoining the hydrologic cycle.

Why is all this so dangerous to the natural environment?

- Fruits and vegetables add multiple excess nutrients to the ocean water, which upsets the nutrient cycle and often contributes to a phenomena called "algae blooms." Algae blooms are the unnaturally rapid reproduction of ocean plankton, which result in a plethora of problems, including but not limited to: 1) formation of large, opaque mats that cover the surface and block sunlight from reaching the ecosystems below the surface; 2) oxygen depletion due to the high quantities of oxygen the algae consume, both while alive and when biodegrading, 3) poisoning the higher life forms, as many of them are toxic and accumulate in the tissue of fish that eat them. Ultimately, the delicate balance of the ocean gets upset and the fishing industry and natural wild-life begin to suffer.

- Soaps and oils are composed of various chemicals depending on their type and use. These chemicals can result in ocean animals suffering from skin diseases, eye problems, contaminated food supplies, birth defects, and death.

- Bacteria can result in worldwide epidemics if left to float around the water system. The World Health Organization estimates that approximately 80% of the sickness, disease, and death in children and infants are directly related to contaminated water.

It's interesting to note that most water pollution problems are actually caused by land based activities as opposed to things like shipping, boating, and fishing.

How do we remove these contaminants from waste water?

- **Primary**: Screen and skim to remove solids and grit; sedimentation (allow solids to settle) may include oil/water separation—created sludge—generally physical.

- **Secondary**: (activated sludge or other—generally chemical).
  - Aeration plus bacteria breaks down organic matter in the liquid,
  - Secondary sedimentation—may include addition of a flocculant (still need to remove nutrients, heavy metals, and some chemical pollutants).
  - Disinfection (addition of chlorine or treatment with UV + ozone)—kill pathologic/bacterial pollutants.

- **Advanced Treatment**: (carbon, sand filters, chemical treatment).

- **Sludge Treatment** (digester + drying beds + re-use as fertilizers): Convert organic matter to stable form, reduce volume, destroy harmful organisms, produce saleable by-products. Sludge treatment, handling, and disposal may account for up to 35 to 50% of the cost of operating a wastewater treatment plant.

OBJECTIVE

The objective of this exercise to undertake the same procedures used in sewage treatment plants and attempt to cleanse water that is "contaminated" with various organics, such that it can be safely released into the environment. Keep in mind that the procedure you are about undergo rids the water of *only* solids and organic contaminants like dirt, leaves, food, and bacteria. In a real sewage plant, you would have to contend with oil, soap, chemicals, and anything else that ends up in the sewer system, all of which require additional treatments in order to remove. Before we begin, locate the data table at the end of the chapter. You will record your data on this table as you and your TA work through the experiment.

Procedure

1. Begin by observing jug #1, which is representative of "raw organic sewage." At this beginning stage, no treatment has been done on this sample, and assuming this water came from a place where there were no chemical contaminants, this is what would be entering the sewage treatment plant. On your data sheet, note the clarity, odors (if any), color, and whether or not solids are present.

2. Next, observe jug #2, which has undergone the first stage of treatment, overnight aeration. Make the same observation as above for this sample on your data sheet.

3. Finally, observe jug #3. This sample has not only been aerated, but has also undergone the second stage in the cleansing process and has been treated with a chemical called potassium aluminum sulfate, or "Alum" for short. Alum is chemical that causes loose particles to mesh together into a jelly-like floc, so that solids separate out from the liquid. Make the same observations as you have for the past two jugs on your data sheet.

4. Take the container labeled "Alum Sample" and fill it up with sample water from jug #3.

5. The next stage in the cleansing process is filtration. Construct a filter by inserting the filter paper into a funnel, then placing a layer of gravel, then medium-grained sand, and finally fine sand in the filter paper as shown. See Figure 12.1.

    a. Run some clean water through the filter prior to putting your sample in.

    b. Find the container labeled "Sand Filter" and place it beneath the funnel.

    c. Pour ¾ of your sample in the "Alum" container through the filter. Make sure to leave about ¼ of the alum sample in the container for comparison purposes.

    d. Make the observations of the sand-filtered sample as you have done previously and record them on your data sheet.

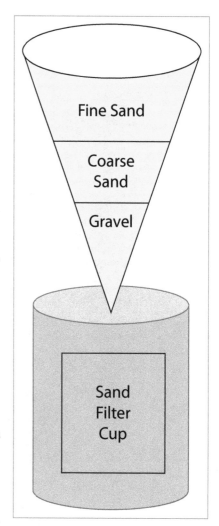

Figure 12.1 - Sand filter

6. Following sand filtration, the sample will need to be filtered through activated carbon. Carbon is exceptionally good at binding with organics (hence the reason biology is carbon-based). As such, this stage in the cleansing process is undergone so the carbon can grab all the dissolved organics in the sample.

    a. Construct a carbon filter by placing a new piece of filter paper in the funnel and filling it with activated carbon.

    b. Flush the filter out by running some fresh water through before putting your sample in.

    c. Find the container labeled "Carbon 1" and place it beneath the filter.

    d. Pour ¾ of your sample from the "Sand Filter" container through the carbon filter. Make sure to leave about ¼ of the sand-filter sample in the container for comparison purposes.

    e. Find the container labeled "Carbon 2" and place it beneath the filter.

    f. Take the sample from the container labeled "Carbon 1" and run it through the carbon filter a second time.

    g. Make the appropriate observations of "Carbon 2" and record your observations on your data sheet.

7. It is now time to disinfect the sample of any bacterial contaminants.

    a. Find the container marked "Bleach."

    b. Pour ½ the sample from the "Carbon 2" container into the container marked "Bleach."

    c. The proper way to proceed at this point would be to pour ½ the sample from the Carbon 2 container into a graduated cylinder, calculate 2% of the total remaining volume of water, and use a pipette to add that amount of bleach. However, because we do not have the appropriate pipettes in the laboratory to measure that quantity of bleach, we will have to approximate the correct quantity, which amounts to 1–2 small drops of bleach from your pipette.

    d. Using the supplied pipette, add 1–2 drops of bleach to the sample in the Bleach container.

    e. Swirl gently for a minute or so.

    f. Make your observations and record them on your data sheet.

8. The final stage is removal of any fine-grained solids that may still be in the sample.

    a. Construct a filter with nothing but filter paper.

    b. Find the container labeled "Final Product" and place it beneath the funnel.

    c. Pour ½ your bleach sample through the filter.

    d. Make your observations and record them on your data sheet.

Data Analysis

Just to get a good look at what you just did, line up the labeled cups side by side in the order they were processed.

Alum → Sand filter → Carbon filter → Bleach → Final product

Using the leftover liquid, take an overall look at the changes that took place at each step.

	Solids	Clarity	Odor	Color
Control				
Aeration				
Alum				
Sand Filter				
Carbon Filter				
Chlorine				
Fine filter				

APPENDIX

All appendix images taken by Robert Agnew, unless otherwise noted.

CHAPTER FOUR

Group A

Group B

Group C

Group D

Group E

Group F

Sample X

Sample Y

Sample Z © PROF. ELLI PAULI

Sample W.1

Sample W.2

CHAPTER FIVE

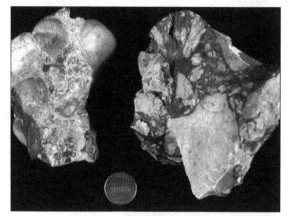

Sample X

Sample Y

Sample Z

Sample 1a

Sample 1b

Sample 1c USGS

Sample 2a

Sample 2b

Sample 2c

Sample 2d

Sample 3

Sample 4

Sample 5 © EURICO ZIMBRES

CHAPTER SEVEN

Sample A

Sample B

Sample C

Sample D

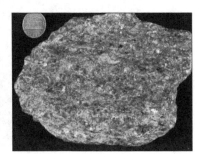

Sample E

Sample F

Sample G

Sample H

Sample I

Samples J and K

Samples L and M

Sample N

Sample O

Sample P   © PROF. ELLI PAULI

Sample Q

Sample R   © PROF. ELLI PAULI

CREDITS